YOU CAN
MAKE A
DIFFERENCE

YOU CAN MAKE A DIFFERENCE

BE ENVIRONMENTALLY RESPONSIBLE

JUDITH GETIS

Boston   Burr Ridge, IL   Dubuque, IA   Madison, WI
New York   San Francisco   St. Louis
Bangkok   Bogotá   Caracas   Lisbon   London   Madrid   Mexico City
Milan   New Delhi   Seoul   Singapore   Sydney   Taipei   Toronto

WCB/McGraw-Hill

A Division of The McGraw·Hill Companies

YOU CAN MAKE A DIFFERENCE: BE ENVIRONMENTALLY RESPONSIBLE,
SECOND EDITION

✪ This book is printed on recycled, acid-free paper containing 10% postconsumer waste.

1 2 3 4 5 6 7 8 9 0 DOC/DOC 9 3 2 1 0 9 8

ISBN 0–07–29416-0

Vice president, editorial director: *Kevin T. Kane*
Publisher: *Edward E. Bartell*
Sponsoring editor: *Daryl Bruflodt*
Marketing manager: *Lisa L. Gottschalk*
Project manager: *Vicki Krug*
Production supervisor: *Sandy Ludovissy*
Designer: *K. Wayne Harms*
Compositor: *Shepherd, Inc.*
Typeface: *10/12 Garamond*
Printer: *R.R. Donnelley & Sons Company/Crawfordsville, IN*

The symbols used are Native American symbols for the earth (circle), land (mountain), water (flowing lines), and air (bird). All are typical of the Western Sioux except the symbol for the earth, which is Western Santee. The symbol for the earth is often seen as a symbol for a whirlwind or the sunburst when not enclosed by a circle. When enclosed, it is a symbol for the earth. The colors in the reproduction on the cover represent the four directions, red being north.

Cover and interior design: *Ben Neff*
Back cover photo by *NASA*

Library of Congress Cataloging-in Publication Data

Getis, Judith, 1938–
    You can make a difference : be environmentally responsible /
Judith Getis. – – 2nd ed.
        p.   cm.
    ISBN 0–07–292416–0
    1. Environmental protection–Citizen participation.
    2. Environmental resposibility.   I. Title.
    TD171.7.G48   1999
    363.7'0525- -dc21

97–46909
CIP

www.mhhe.com

CONTENTS

INTRODUCTION

On Board Spaceship Earth

Global Warming Calamity

Toxic Waste Crisis

Pollution Forces Beach Closure

Death of a Rain Forest

Hole in the Ozone Layer

It seems as though we encounter newspaper headlines and magazine articles like these wherever we turn. Radio and television, too, carry special reports about environmental problems. Is it all hysteria, or is there a real cause for concern?

In this case, the old adage holds true: Where there's smoke, there's fire. We have been abusing the earth for centuries. As population numbers have soared, and as industrialization has become a worldwide phenomenon, we have begun to recognize the deadly consequences of environmental abuse. It isn't simply a matter of aesthetics or the quality of life, though those are important concerns. Nor is it just a matter of the extinction of some rare and endangered plant or animal species—though such extinctions will also affect us in ways we haven't even dreamed of yet. No, it is a fact that our actions, and those of the generations who have gone before, are threatening our own health and safety. Dirty air kills people. So do dirty water and radioactive wastes. It is time for us to learn how to live in harmony with the environment.

Spaceship Earth

Look at the earth in the photo on the back cover. It's just a small blue dot hurtling through space, one of billions in the galaxy. Think of it as a spaceship—Spaceship Earth. We're on board for a very long journey, one that will last our lifetimes. The spaceship contains everything we need in order to live, but we can never take on new supplies.

To survive, we need to use the resources Spaceship Earth has to offer, but we must do so wisely and carefully. That is, we must balance the ability of the earth to support life with the demands we make on it. We must recycle everything we need in order to live: water, air, and essential chemicals and nutrients. That recycling capability is our life-support system. Without it, we'd soon exhaust our resources.

There is no place to throw anything away, of course. "Away" is always with us. Even if we bury things in the ground or dump them in the ocean, they're still on board Spaceship Earth. Fortunately, wastes can be renewed by natural cycles as long as we don't overload them or burden them with substances that nature cannot break down.

Even the amount of room is finite. We can't make Spaceship Earth any larger than it is. That means that we can't pollute, ravage, or destroy one part of the spaceship and then move on to unspoiled areas indefinitely. Not only does the destruction of one area affect the rest of the spaceship, but at some point we'll run out of room.

The Biosphere

Spaceship Earth is just an analogy, but it's a useful one if it reminds us how finite and fragile our environment is. All life exists within a thin spherical film of land, water, and air that is called the **biosphere.** The biosphere is composed of three interrelated parts: the lithosphere, the hydrosphere, and the atmosphere, or, as they are more commonly known, the land, air, and water.

The **lithosphere** is the top portion of the earth's crust. Only a few thousand feet deep, it contains the soils that support plant life, the minerals that plants and animals require to exist, and the ores and fossil fuels (coal, oil, and natural gas) that people exploit.

The **hydrosphere** is our water supply. It consists of surface and subsurface waters in oceans, rivers, and lakes, as well as groundwater. Much of it is locked in ice or earth and not immediately available for use.

The **atmosphere** is an envelope of usable air, only 7 miles (11.25 km) high, that surrounds the lithosphere and hydrosphere.

The biosphere is an intricately interlocked system containing all that is needed for life, all that is available for life to use, and all that *ever* will be available. Because everything is related to everything else, everything is interdependent. The ingredients of the thin biosphere must be, and are, constantly recycled and renewed in nature. Plants purify the air; the air helps purify the water; the water and the minerals are used by plants and animals and then returned for reuse.

The biosphere provides us with what we need, but it also obligates us to certain actions in return. Because in the past the supplies of essential resources like food, air, and water generally were greater than the demands placed upon them, we have become complacent—lulled into thinking that we're not indebted to the earth for what it gives us. It's up to us to find out how to repair the damage that we've done and how to protect the environment from further harm.

This Handbook: Challenge and Response

This book is organized around the three parts of the biosphere: land, water, and air. Each section contains descriptions of the environmental problems associated with that part of the biosphere. In the section on the atmosphere, for instance, we discuss such problems as acid rain and the greenhouse effect. Immediately following each problem or "Challenge," we suggest ways that individuals can help solve or alleviate it—i.e., the "Response."

We've called the book *You Can Make a Difference* because we really believe that. Every person counts. Everything you do and everything you buy has some effect on the environment. And everything you do has some impact on other people. You needn't become an activist, attending demonstrations or writing letters to the city council, in order to influence others. You're doing that simply by living.

THE LAND

CHALLENGE: THE GARBAGE MESS

If we imagine ourselves aboard a spaceship, we realize immediately that one of the first problems we'd encounter would be how to dispose of trash. Garbage won't simply go away. Any waste we create has to be either stowed somewhere or converted into a usable product. If we earth dwellers had been clever and had thought about the problem beforehand, we'd have seen to it that we created as little trash as possible and that most of what we did produce was capable of being reused, recycled, or recovered.

Until recently, North Americans didn't have to worry too much about the trash they produced. There was plenty of open space in which to dump it, and there wasn't as much trash even a generation ago as there is today. Furthermore, most of it was biodegradable (it would decay naturally) and not nearly as hazardous as it is now.

American communities are currently facing twin problems in disposing of their wastes: the sheer volume of the trash and the toxic nature of much of it. In the United States and Canada combined, the amount of trash has doubled in the last thirty years and currently stands at some 200 million tons per year, or about 3.5 pounds (1.6 kg) per person per day. You might find these figures hard to believe, arguing that *you* certainly don't throw out that much. But you have to remember that municipal waste includes things like rusted cars and broken refrigerators as well as newspapers and beer cans. (Municipal waste does *not* include waste from manufacturing and industrial processes. That's another problem societies have to deal with.)

About three-fourths of the trash is deposited in so-called sanitary landfills—dumps where each day's waste is compacted and covered by a layer of clean earth. "Sanitary" is a deceptive word. Rain washes toxic chemicals from paint, pesticides, and other products into the soil, from which it eventually seeps into groundwater supplies. Decomposing food wastes produce methane, a volatile gas that leaks into the air and soil.

Municipal landfill capacity is shrinking dramatically. The number of landfills in the United States fell from 18,000 in the late 1970s to 4,000 in 1996. Many of these will be closed soon, either because they are full or because they are a danger to the environment. Facing an acute shortage of landfill space, some northeastern states are paying other states and even other countries to accept their trash. As people have become more aware of how landfills contaminate air, water, and soil, it becomes increasingly difficult for communities to site new landfills. Everyone wants them, but "NIMBY"—not in my backyard.

One alternative to burying trash is to burn it. About 15% of America's garbage is burned in incinerators that use extra-high temperatures to reduce trash to ash. Incinerators pose environmental problems of their own by generating toxic pollutants. Air emissions from incinerator stacks contain an alphabet soup of highly toxic elements, ranging from A (arsenic) to Z (zinc). The ashes left after burning, which are also toxic, are buried in landfills.

The "Five R's"

If we can't safely bury or burn our trash, how is it best handled? An easy way to think about the problem of waste management is to focus on the 5 R's: **reduce, reuse, recycle, reject,** and **reward.** These are the solutions to trash disposal. Practical steps you can take to implement them are outlined in the next several pages.

Response: Reduce the Amount of Trash

If we produce less waste in the first place, we will shrink the amount of trash needing to be disposed of. Manufacturers can reduce the amount of paper, plastic, glass, and metal they use to package food and consumer products. Over the last 30 years, for example, the weight of plastic soft-drink bottles and of aluminum cans has been reduced by 20–30%. Some products, such as detergents and beverages, can be produced in concentrated form and packaged in smaller containers.

Here are some ways you can do your part. You probably won't implement all of them, but even adopting just a few will help cut down on the amount of trash you throw away.

Whenever possible, buy durable rather than disposable items.

- Use cloth rather than paper napkins.

- Use sponges or a cloth towel rather than paper towels to wipe up spills.

- Use newspapers instead of paper towels to clean your windows *if* they are the kind that won't leave inky smudges.

- Store food in containers rather than covering it with aluminum foil or plastic wrap. If you must use a wrap, try cellulose or waxed paper, which are biodegradable.

- Use a lunch box or small tote bag instead of a paper bag for your lunch, and a thermos bottle instead of a single-serving container of juice or milk.

- For your morning cup of coffee and doughnut, use a ceramic coffee mug and plate, not a paper or styrofoam cup and plate. Use metal instead of plastic cutlery. It's foolish to throw away something you've used for just a few minutes, when it will take decades or centuries to decompose in a landfill.

- If you're having a party, borrow glasses and dishes from a friend rather than use disposable ones.

Containers and packaging waste make up about one-third of the average person's trash. Try to avoid buying overpackaged products. By "overpackaged" we mean those that are needlessly packaged, those that are packaged in nonbiodegradable material, like foam or plastic, or those packaged in single-serving units.

- A classic example of needless packaging is oranges on a styrofoam tray with a plastic overwrap. Whenever possible, buy loose fruits and vegetables that you bag yourself instead of prepackaged produce. Many supermarkets have bins of nuts, candies, dried fruit, trail mix, and the like. You can help yourself to the amount you need and avoid the excess prepackaging.

- If you need nails, washers, fuses, or other such items, patronize a hardware store that has them in bulk. You'll be able to get the ten nails you need instead of twenty that are packaged in cardboard and plastic.

- Whenever there is a choice, avoid products packaged in nonbiodegradable material. Thus you should buy butter or oleo that comes wrapped in paper rather than in plastic tubs. Buy bars of soap wrapped in paper, not liquid soap in a plastic bottle. Buy soft drinks that come in returnable or recyclable containers, preferably bottles or cans.

- Try not to buy products packaged in single-serving units. This includes individually wrapped slices of cheese or single-serving packages of milk, juice, pudding, raisins, cookies, and crackers. In general, buy nonperishable food items in the largest size that you can afford and can store. You'll reduce the amount of packaging waste and also save money.

You may not know this, but you can stop most junk mail from arriving in your mailbox. Junk mail is a tremendous waste of resources—millions of trees annually, as well as the energy required to print and distribute the mail and then to cart it off to a landfill or incinerator where, once it's disposed of, it will contribute to pollution of the ground, water, and air. Contact the following organizations and ask that your name be deleted from junk mailing lists. To make sure most of the mail stops coming, include all the variations of your name as it appears on mail.

- Mail Preference Service, Direct Marketing Association, P.O. Box 9008, Farmingdale, NY 11735-9008.

- Direct Marketing Association Telephone Preference Service, P.O. Box 9014, Farmingdale, NY 11735-9014.

- Private Citizen, (800) 288-5865.

- Zero Junk Mail, (888) 665-8656.

- You can also tear off the mailing label on any advertising mail you receive but don't want and send it to the company along with a note asking to be removed from its list. The label helps the company find your name in its computer file.

Response: Reuse Items Whenever Possible

Reusing things however and whenever feasible cuts down on the amount of trash going to landfills or incinerators. It also reduces the amount of energy used to produce goods, and it will save you money because you won't have to buy as much. Here are some suggestions for reusing things:

- Paper and plastic bags can be reused a number of times before they need to be discarded. Paper bags are good wastepaper basket liners; plastic bags from the produce section of the supermarket are ideal for covering food in the refrigerator.

- Use plastic bags from the dry cleaners to cover out-of-season clothes in your closet, to line your suitcases, or to wrap Christmas ornaments.

- Use plastic food containers to store food or items in your desk or workshop.

- Strips cut from worn-out nylon stockings are perfect for tying plants to stakes to hold them upright.

- Reuse gift wrap, boxes, ribbons, and bows. Wrap a child's present in the Sunday comics.

- Use rechargeable batteries in your radio, flashlight, and camera.

- Save ashes from the barbecue grill and fireplace in a trash can, then sprinkle them on icy sidewalks and driveways in the winter time.

If you no longer have any use for an item, think of an organization that could use it. Donate used clothing to a charity or church for its bazaar. Sell used books to a dealer in secondhand books or donate them to a public library. Give magazines to a friend or to a hospital or retirement home. Eyeglasses can be donated to an organization that will distribute them to needy people. Donate used furniture and appliances to a daycare center, a shelter, or a charity. If you live in a group setting, such as a dormitory or apartment house, it would be efficient to collect donations from a number of people. If you have enough material, an organization such as the Salvation Army or St. Vincent de Paul will pick it up.

Response: Recycle Cans

The trash we bury in landfills represents huge amounts of wasted energy and wasted resources. We annually bury enough aluminum to rebuild the entire American commercial airfleet. The wood and paper dumped in landfills could heat at least 5 million homes for two hundred years.

Recycling reduces the amount of waste needing disposal by making a portion of it available for reuse; it is turned back into "raw" materials for manufacturing. Americans recycle only 13% of their waste, but analysts say up to 50% is recyclable.

The ways you'll be able to recycle will depend on the kinds of recycling services available where you live. Some schools and some municipalities make it easy to recycle. Bins or dumpsters may be available in your dormitory or apartment house. (If they aren't, you might talk to the management about setting up a recycling program for the building.) There may be collection sites in supermarket or shopping center parking lots. Your community may even have curbside pickup. If you're not sure what recycling programs are available in your area, call the local sanitation department or a commercial recycling center.

Aluminum cans are easily recycled. Most communities have numerous places to drop off cans, and an increasing number have curbside pickups of recyclables. Some shopping centers have reverse vending machines that will pay you for aluminum cans.

Three Good Reasons to Recycle Aluminum Cans

1. An aluminum can thrown away in the trash will still be in the landfill hundreds of years from now. A can that is returned and recycled is typically back on the supermarket shelf within two months.

2. Many states now require a refundable deposit on beverage containers. You'll get that money back if you recycle the can.

3. Making aluminum from raw materials is expensive because of the electricity required in the manufacturing process. But it takes only 5% to 10% as much electricity to make aluminum from scrap. The water and air pollution that accompany manufacturing are also reduced by 90% to 95%.

Incidentally, aluminum in other forms is also recyclable. This includes aluminum foil and frozen food trays, aluminum siding, aluminum pots and pans, and even lawn furniture. All of these can be made into similar products or into castings for new automobile parts. You would probably have to take such items to a recycling center.

Tin Can Recycling   All tin cans (which are really steel cans coated with either tin or chromium to prevent rusting) are also recyclable. Steel producers use scrap to make a variety of products. It is cheaper than making steel from scratch; it reduces pollution; and recycling keeps the cans out of the landfill.

Response: Recycle Glass Bottles and Jars

Americans annually throw away almost as many glass jars and bottles as they do aluminum cans—roughly 30 billion a year. According to the Environmental Defense Fund, that's enough glass to fill the 1,350-foot (412 m) twin towers of New York City's World Trade Center *every two weeks*. Bottles and jars take up space in landfills needlessly, for almost all are recyclable. Manufacturing new ones to replace those that are discarded involves mining white sand, limestone, soda ash, and feldspar, transporting these materials to the plant, dissolving them by superheating, then cooling the mixture. These processes require energy, produce mining waste, and contribute to air and water pollution.

All glass bottles and jars are recyclable. Clear glass is used for some juice and beverage containers, peanut butter and jelly jars, and a variety of other food containers. Apple juice, prune juice, and beer often come in brown bottles, while green bottles are used for wine, beer, and a few soft drinks.

To reduce their volume, recycled glass jars and bottles are broken up before being shipped to the manufacturing plant. This recycled crushed glass is called *cullet,* and glass containers made in the United States typically contain a percentage of cullet. Manufacturers like it because it saves them money. Cullet lowers the temperature required to melt the sand-lime-soda mixture, so that less energy is consumed in the manufacturing process. And for every ton of cullet used to make new containers, approximately 1.3 tons of raw materials can remain in the earth. Specifically, for every ton of glass that is recycled, we save 1330 pounds of sand, 433 pounds of limestone, 433 pounds of soda ash, and 151 pounds of feldspar.

Some kinds of glass are *not* recyclable: drinking glasses, plate glass, mirrors, Pyrex, crystal, and light bulbs. These cannot be combined with cullet from glass jars and bottles because they don't melt at the same temperature.

Recycling glass is almost as easy as recycling aluminum cans. Remove the caps and plastic rings, but don't bother with paper labels. Until you're ready to recycle, store the glass in a separate container from landfill-bound trash. (If space is a problem, you can combine all your recyclables in one large container and separate them when you get to the recycling center.) In states that have "bottle bills," you can return glass soft drink bottles to the store where you purchased them for a refund.

Response: Recycle Newspapers

Paper constitutes the largest portion of our trash, and newspapers alone take up 15% of the space in a typical landfill. One of the problems with newspapers is that they don't decompose much in landfills that have liners intended to prevent toxic materials from leaching into the soil. Another problem is that when they do decay, the inks leach into the soil and groundwater.

A decade ago, few Americans recycled their newspapers. Now the figure is up to one-third, thanks to states and communities that have passed mandatory recycling laws and to increasing awareness of the benefits of recycling.

- Recycling saves trees. Recycling one ton of newspaper saves seventeen trees from being ground up into virgin wood pulp. The print run of a single Sunday edition of *The New York Times* takes 75,000 trees. The newsprint for a single edition of all the Sunday papers in North America takes more than a half-million trees.

- Much less energy (about 30% to 50%) is required to make new paper from old paper than to make new paper from scratch.

- Recycling saves over half the water needed to manufacture virgin paper. Recycling just one ton of newspapers conserves 7,000 gallons of water and also reduces the water pollution that results from the production of virgin paper.

- Recycling that ton of newspapers also keeps some 60 pounds (27 kg) of air pollutants from being discharged into the atmosphere.

- Finally, recycling helps preserve the ecological diversity of our forests. Although timber companies replant the areas they log, the new forests don't resemble the original. They are really tree farms with a very few species of trees arranged in orderly rows. They lack the undergrowth of flowers, vines, and shrubs that characterize older forests and offer food and shelter for wildlife.

You may not get paid directly for the newspapers you recycle, but you'll be doing a world of good.

To recycle, stack the newspapers inside a brown paper supermarket bag or tie them with string. Do not include the Sunday magazine section or advertising circulars that are printed on glossy paper. Drop the papers in the appropriate receptacles in supermarket or shopping center parking lots, take them to a commercial recycling center, or put them out for curbside pickup.

Response: Recycle Some Plastics

Plastics constitute a small but rapidly growing percentage of trash. North Americans throw away about 60 million plastic bottles every day, as well as thousands of tons of plastic in other forms—garbage bags, food wrappers, and so on. Plastics don't decompose in landfills. If they are incinerated, they release chemicals into the air that pose a threat to human health and the environment. Plastic six-pack holders, plastic bags, and other plastic litter blown into the water from landfills are deadly. Each year well over a million birds and fish die from starvation or choking when they ingest or get tangled in such debris.

Because there are many different kinds of plastic, recycling is a problem. Not all plastics can be melted down and recast. In fact, technically, plastics are reprocessed rather than recycled. Because plastic degrades when melted, it cannot be recycled into the same product as the original. Thus, soda bottles may become traffic barrier cones or fiberfill for sleeping bags, but not new soda bottles.

Seven types of plastics are commonly used. The code number stamped inside a recycling symbol (arrows chasing arrows) on the bottom of each container indicates the type of plastic.

1   PET or PETE (polyethylene terephthalate) is the plastic used for soft drink bottles, liquor bottles, peanut butter and mustard jars, and so on. Although recycled PET is not used to make new soda bottles, it can be used as fiber for carpets and filler for sleeping bags, ski jackets and polyester clothes, and a variety of other products.

2   HDPE (high-density polyethylene) is used for milk jugs, detergent and bleach bottles, butter tubs, motor oil containers, and other items. It can be turned into items such as trash cans, flowerpots, toys, and lumber for park benches and boat piers.

3   PVC (polyvinyl chloride) is found in some shampoo, cooking oil, water bottles, and other similar products, but most of it is used for construction and in plumbing. It is not normally recycled.

4   LDPE (low-density polyethylene) is used for sandwich and produce bags, as well as wrap for dry cleaning. Some supermarkets have bag-recycling bins.

5   PP (polypropylene) is a plastic used in many diverse products, such as food lids, bottle caps, yogurt and cottage cheese containers, disposable diapers, and encasements for car batteries. It is rarely recycled.

6   PS (polystyrene and polystyrene foam) is the light, white plastic used to manufacture disposable foam cups, plates, meat trays, burger boxes, and packing "peanuts."

7   denotes mixed resins. Because they are made from more than one type of plastic, containers such as squeezable catsup, jelly, and fudge bottles cannot be recycled.

At present, only PET and HDPE containers are frequently recycled, but every community is different. Ask your local recycling center for the types of plastic it accepts.

Response: Recycle Yard and Food Wastes

Depending on the season and the part of the country where you live, anywhere from 15% to 35% of the garbage in your local landfill probably consists of food and yard wastes. This includes things like banana peels and coffee grounds as well as dead leaves and grass clippings. These wastes are all organic, which means they will decay naturally if exposed to air and water. Burying them in landfills, however, slows down the rate of decomposition tremendously and also takes up valuable landfill space. Fortunately, recycling this waste is relatively easy.

Many communities have established composting programs in order to reduce the volume of yard waste that is landfilled. The waste is collected separately from other wastes, then composted. Composting is the process of letting millions of live organisms "eat" organic material and turn it into a nutrient-rich, crumbly soil called *humus*.

If there is no municipal composting program where you live, you can recycle the yard and food wastes yourself by making your own compost pile.

What Can Be Composted?

- Grass clippings, dead leaves, twigs, and shredded branches are all compostable. Don't put weeds or diseased plants on the pile.
- To the yard waste, you can add all fruit and vegetable peelings, all grains, eggs, and even tea bags and coffee grounds. Decomposition will be quicker if you break up any large pieces, such as watermelon rinds or grapefruit skins. It's best not to add meat, bones, fat, or grease because they take too long to break down and attract dogs, cats, and rats and other rodents.

How to Compost   You can buy a compost bin at your garden store or build a simple one yourself in a corner of the yard. It can be made of scrap lumber, concrete blocks, or even chicken wire. It should be from 3 to 5 feet (1 to 1.5 m) high and have adequate ventilation. Toss food and garden waste into the bin and occasionally add a layer of soil or bonemeal. Keep the mixture moist and turn it periodically to allow the air to circulate. In one to four months, the compost should be dark brown and crumbly, ready to be used in your garden. Your local garden store will have advice and supplies.

Benefits of Compost   Compost can be dug into the soil in your garden or used as a mulch around trees and shrubs. Mixed with potting soil, it is excellent for houseplants or starting seedlings. Some of the benefits of compost are as follows:

- The decomposed material is rich in nutrients.

- The humus improves soil structure and texture. It helps hold sandy soil together and helps break up clay soils, making it easier for plant roots to penetrate the soil.

- It increases the soil's ability to hold water, cutting down on water loss due to evaporation during hot summer months.

Response: Reject Disposables and Nonbiodegradables

Reduce, reuse, recycle. A fourth "R" is reject. We suggest here that you reject certain kinds of products when you go to the store: disposables and nonbiodegradables. You can help solve the garbage crisis by not buying single-use, so-called disposable items like diapers, razors, cameras, and flashlights. Perfectly good (and less expensive) alternatives exist for all of these products—ones that can be used over and over again before they need to be discarded.

Of all the disposables, diapers present the largest problem. A child will wear 5,000 or more disposable diapers before being toilet-trained; a child wearing cloth will use about fifty diapers. This accounts for the fact that about 20 billion disposable diapers are used each year in the United States and Canada. They are made from over one million tons of tree pulp, the equivalent of about 20 million trees per year. Just their manufacture results in millions of pounds of pollutants being spewed into the air.

Disposable diapers present three main problems:

1. By volume, they account for about 3% of the waste going to landfills, where they take up space that could be used for other things.
2. The plastic part of the diapers doesn't decompose readily in a landfill.
3. Most people don't dunk disposable diapers in a toilet to rinse out the waste, so they are still filled with urine and feces when they end up in the landfill. Human biological waste isn't supposed to be put in landfills because contaminants will slowly leach into the earth and pollute groundwater supplies.

Despite these facts, most parents will opt to use disposable diapers instead of buying and washing their own cloth diapers or using a diaper service. If you are among them, don't be too concerned—your choice won't have a major impact on the health of the environment. Energy and resources are used to manufacture cloth diapers, and washing them uses more water and energy.

"Degradable" Plastic   What about diapers or other plastics (trash, grocery, and merchandise bags) now being marketed as "degradable" and "safe for the environment"? Don't be fooled. A truly degradable plastic isn't yet on the market.

Some plastics are called "photodegradable." They have to be exposed to sunlight in order to break down—but garbage in landfills isn't exposed to sunlight.

Others, labeled "biodegradable," break down in the presence of bacteria. They contain an additive, usually cornstarch, to allow them to degenerate, but it constitutes only 6% to 10% of the item's volume. Once the bacteria in the soil eat the starch, the plastic degenerates into many little plastic pieces, which can be carried away from the landfill and deposited in streams and oceans. Furthermore, extra plastic has been added to the item in the first place to compensate for the weakening effect of the cornstarch, so the total volume of plastic reaching the landfill hasn't been reduced at all.

A truly degradable plastic, one that degrades completely and harmlessly, will have to be made entirely from starch. Until such a product is developed, you're not doing the environment a favor by buying so-called degradable plastics.

Response: Reject Polystyrene Foam Products

Better known under its Dow Chemical Company brand name, Styrofoam, polystyrene foam is a form of plastic used to make hundreds of products. You are probably most familiar with it in food containers—cups, plates, egg cartons, trays for meat and produce, ice chests, and so on. It is also used to make insulation, flotation devices, foam cushions, and packaging (both packing pellets and form-fitting containers for electronic goods).

Until 1988, all polystyrene foam was made with chlorofluorocarbons, or CFCs, which are a major contributor to the depletion of the ozone layer (see pages 76–77). Some types of polystyrene foam are still made with CFCs, but the food industry no longer uses them to manufacture foam containers. Many are now made from HCFC-22. It has only one-twentieth the ozone-depleting potential of the original compound, but it still poses a threat to the ozone layer.

There are many good reasons to avoid buying and using polystyrene foam:

- Like other plastics, polystyrene foam is a petroleum-based product, and its manufacture depletes the earth's limited supply of oil.
- When flexible foam is produced, CFCs are released into the atmosphere. With rigid foam, the CFCs are trapped in the foam cells. They are released into the atmosphere when the foam is broken or crushed in a landfill. Incinerating polystyrene produces toxic fumes.
- Polystyrene foam is not biodegradable. Unless it is burned, the coffee cup you threw out this morning will remain in the environment forever.
- The technology for recycling the foam is still in its infancy. Until it is better developed, those cups and plates are not recyclable.

The bottom line is as follows:

- Don't buy polystyrene foam containers, like ice chests or cups. If you must use disposable cups and plates, buy paper products.
- Don't buy products that come in polystyrene containers. Buy eggs in cardboard, not foam, cartons. Patronize fast-food restaurants that have switched to paper packaging.
- Don't buy foam pellets for shipping fragile items. Crumpled newspapers or straw work just as well.
- Don't buy mattresses or furniture with foam cushions unless they are labelled "Non-CFC."

Response: Reward Companies That Demonstrate Concern for the Environment

Whenever you purchase something, whether it's a birthday card or an automobile, your choice will have some effect on the environment. Is the card made from recycled paper; is it printed on bleached paper; does it contain plastic parts or toxic dyes? How fuel-efficient is the car; does it have an overdrive gear; is it equipped with air conditioning or power windows? Answers to questions such as these help determine the environmental impact of a product.

You can be a "green" consumer—one who buys goods and services that in their manufacture, use, and disposal are least harmful to the environment. Support companies that manufacture environmentally friendly products, that minimize waste, and that practice and encourage recycling.

Ample evidence indicates that corporations are becoming more sensitive about how their goods and services affect the environment *and* about how their companies are perceived by their customers. If enough consumers show by their purchases that they want to buy environmentally benign products, companies will pay attention. Public opposition to fishing techniques that trapped dolphins in tuna drift nets led the three biggest tuna-canning companies to agree to buy only from fishing boats that use other methods. McDonalds changed from styrofoam to paper packaging because a few of its customers spoke up; other fast-food chains soon followed suit.

How can you identify which products in a certain category are environmentally preferable? The Canadian government has established a program called Environmental Choice. Products that meet certain standards are identified with a symbol of three intertwined doves. The products must be derived from renewable resources; they must be nontoxic and either recyclable or biodegradable; and their packaging must meet certain criteria.

In addition, Canada's largest supermarket chain, Loblaws, carries a line of President's Choice GREEN products. Items with this stamp of approval include unbleached, reusable coffee filters and toilet tissue made from recycled paper.

Identifying environmentally friendly products in the United States is more difficult. At present there are no nationally accepted standards for determining what products are environmentally sound—or even standards that define such terms as "biodegradable" and "recyclable." Even the recycling logo (arrows chasing arrows) is confusing. It means either that the product is recyclable or that it is made from recycled materials. Paper companies, however, make a distinction between the two. Three arrows on a dark background indicates that the paper contains recycled fibers. Three arrows with no background simply means the paper is recyclable.

Recognizing the marketing importance of environmentalism, a few companies have attempted to appear environmentally sensitive by making irrelevant, deceptive, and even false claims about their products. An aerosol spray can may proclaim "No CFCs" when it still contains other hazardous compounds. A plastic trash bag might be touted as "photobiodegradable" in spite of the fact that state landfill regulations will not permit it to be exposed to the elements so that it can decompose. An environmentally harmful product, such as a high-phosphate detergent, may come wrapped in "green" packaging made from recycled paper.

One U.S. retailer, Wal-Mart, identifies what it considers to be environmentally preferable products with green and white labels that point out the products' beneficial features. The store determines which items merit recognition by considering both their composition and packaging.

Another source of environmentally sensitive products are mail-order companies. Among them are the following:

- Earth Care Paper Co., PO Box 7070, Madison, WI 53707 (608-277-2900). Sells recycled paper goods, including greeting cards, gift wrap, and computer paper.

- Greenpeace, c/o Prime Time Marketing, 410 Townsend St., Suite 100, San Francisco, CA 94107 (800-916-1616). Sells a variety of goods, including clothing, books, and solar watches.

- Real Goods, 966 Mazzoni St., Ukiah, CA 95482 (800-762-7325). Offers energy-efficient products, from low-flow shower heads to compact fluorescent lightbulbs.

- Seventh Generation, Colchester, VT 05446 (800-456-1177). Has a wide variety of household products, from cellulose sandwich bags to safe cleaning products.

Finally, several books review products and how they affect the environment. Among them are the following:

- *The Canadian Green Consumer Guide* (Toronto: McClelland & Stewart, 1989).

- *Ecologue: The Environmental Catalogue and Consumer's Guide for a Safe Earth*, ed. Bruce Anderson (New York: Prentice Hall Press, 1990).

- *The Green Consumer*, John Elkington et al. (New York: Penguin Books, 1990).

- *Mother Nature's Shopping List: A Buying Guide for Environmentally Concerned Consumers*, Michael D. Shook (New York: Citadel Press, 1995).

- *Shopping for a Better Environment*, Lawrence Tasaday (Deephaven, MN: Meadowbrook Press, 1991).

- *Shopping for a Better World*, Ben Hollister et al. (New York: Council on Economic Priorities, 1992).

Response: Purchase Recycled Goods

Whenever possible, buy products made from recycled material. You begin the recycling process when you turn in containers, newspapers, and so on. Now "close the loop" by buying recycled goods. Recycling won't work unless there is a market for products made from recycled materials. You can choose from thousands of recycled products that are made in North America.

Supermarkets carry goods made from recycled paper, including facial tissue, toilet tissue, napkins, paper plates, paper towels, and coffee filters. Most aluminum cans and glass bottles contain some recycled material.

Office supplies made from recycled paper products include copy paper, cards, envelopes, folders, notebooks, binders, and cardboard boxes. In addition, you can find recycled computer disks, recycled toner cartridges for photocopiers and laser printers, and items such as pens, rulers, and scissors made from recycled plastic.

Hardware stores carry products such as trash cans, hoses, door mats, and floor tiles made from recycled rubber and plastic.

Automotive products include recycled oil, recycled car batteries, and retreaded tires.

Clothing that is "recycled" (secondhand) can be purchased at "nearly new" shops, thrift shops, and garage sales. A few clothing manufacturers, such as Patagonia, make clothes partly from recycled plastics. The Deja Shoe Company makes shoe soles from recycled tires.

What about Cost?   Many recycled products cost less or no more than those made from virgin materials. When they cost more, it's usually because the market for them is limited. The prices will come down when people begin buying recycled goods on a regular basis.

Response: Give Environmentally Responsible Presents

We're often perplexed about what to give a friend or family member for birthdays and holidays. Here are some suggestions for gifts that won't harm the environment:

- Membership in an environmental or conservation organization such as Greenpeace, the National Audubon Society, the Nature Conservancy, or the World Wildlife Fund. There are many others. Their catalogues contain numerous gift items, including T-shirts, calendars, tote bags, games, and toys. Proceeds from the sale of these items help fund the organizations' projects.

- Membership in a museum.

- Tickets to a movie, play, musical, or sporting event.

- A service that you perform yourself, such as babysitting, cleaning the car, putting up storm windows, cooking, or addressing holiday greeting cards.

- A book such as *50 Simple Things You Can Do To Save The Earth, Save Our Planet,* or *Shopping for a Better World.*

- *Gifts That Save the Animals,* by Ellen Berry, lists more than 1,000 items sold by nonprofit organizations that are committed to protecting animals.

- A yard composter that you build or buy.

- A house plant, shrub, or tree.

- Brazil nuts, cashews, other tropical nuts or candy made from them. (Buying these products provides an economic incentive to the people now living in endangered tropical rain forests. For more information, see page 35.)

- Stationery, birthday and holiday greeting cards, notepads, or wrapping paper made from recycled paper.

- An aluminum can crusher, which turns aluminum cans into flat discs that can be stored easily until they can be recycled.

- An electric or solar-powered battery charger.

- A solar-powered calculator, radio, watch, flashlight, or lantern. For the biker or jogger, a solar-powered light to strap on the arm or leg.

CHALLENGE: HAZARDOUS WASTE

You probably associate hazardous waste with environmental disasters like Love Canal or Times Beach, Missouri, or perhaps with radioactive waste from nuclear power plants. You might be surprised to learn that hazardous materials are found in virtually every home in America, and that the typical American throws out 10 pounds (5 kg) of hazardous waste a year.

Hazardous waste is defined as discarded material that may pose a health and safety threat to humans, wildlife, or to the environment when it is improperly stored, transported, or disposed of. Household products contain toxic chemicals that range from A (aldicarb, used in pesticides) to Z (zinc, used in batteries). These products can be found in every room of a house, for they include oven and drain cleaners, furniture polish, used motor oil, and garden weed killers and pesticides.

The United States generates about 275 million tons of hazardous waste a year, or about one ton (2,000 lb) per person. Canada produces some 6 million tons annually. That means this year, next year, and so on. Think how that will add up over your lifetime. Industries, agriculture, and power plants are responsible for most of that waste, but everyone who uses the products the industries have made, or the energy produced by the power plants, contributes indirectly to the waste.

Hazardous wastes can be liquids, gases, or solids. They may be hazardous for a number of reasons.

- **Poisonous** or **toxic** substances can be lethal to humans, animals, and plants. If exposure is slight, you might experience nothing more than a sore throat, dizziness, or nausea. But prolonged exposure to toxic substances can increase your risk of developing cancer, leukemia, and respiratory illnesses. Some toxics are suspected of causing reproductive problems and birth defects.

- **Corrosive** or **caustic** substances, such as battery acid or drain cleaner, burn or eat away at other materials, including flesh.

- **Flammable** materials, such as gasoline or solvents, can catch fire.

- **Explosive** or **reactive** substances (e.g., bleach and ammonia) can explode or release toxic vapors when they are mixed.

How do you know if something you buy is hazardous? The boldfaced words in the preceding paragraph are a good indication. If they appear on the label, the product is hazardous. Other warning words that might appear are *Poison, Danger, Warning,* or *Caution.* Poison and Danger are the worst; they mean the contents are highly toxic.

Unfortunately, about 100,000 chemicals are currently in commercial use, and new ones are being developed every year. We tend to take their safety for granted, but in fact most have not been properly tested for their potential to cause adverse health effects in the long run. For more than 80% of them, there is no toxicity information whatever.

Let us assume that you have a hazardous product and want to get rid of it—used motor oil or half a can of drain cleaner. This is what you should *not* do:

- Pour it into the sink or toilet. From there it will go into the city water treatment system, which is incapable of neutralizing it. Furthermore, it can release toxic fumes and is likely to contaminate the water into which the sewage is eventually discharged.

- Pour it onto the ground or into a storm sewer. It will be washed into a stream or filter down through the earth into an underground water supply.

- Put it out with your regular trash. It will either be incinerated (burned), in which case it will release toxic gases into the air, or be buried in a landfill. Then the can will open when a bulldozer runs over it, and the oil or drain cleaner will leak out and eventually contaminate the soil or water supply. Household toxins have started landfill fires that released toxic fumes across neighborhoods, and they have exploded, injuring or killing landfill workers.

Although these are the ways one should not dispose of hazardous waste, they are the methods most people use. We tend to assume that hazardous waste disposal is a problem for the industries that generate the waste, and we aren't aware of which household products are hazardous. In the pages that follow, we indicate how to handle hazardous materials and suggest some safe alternatives.

Response: Handle, Store, and Dispose of Hazardous Wastes Properly

If you must not pour hazardous wastes down the drain or on the ground, or even put them out with your regular trash, what on earth should you do with them? Recall the Spaceship Earth analogy, then see if the following steps make sense.

Learn to Recognize What Materials Are Hazardous   If you think a product you're considering buying might be hazardous, read the label. Look for words such as "danger," "warning," or "caution." Some labels describe the hazard with words like "corrosive," "flammable," "toxic," or "explosive." And sometimes labels indicate a health risk, e.g., "irritant," or "harmful if swallowed."

Purchase the Least Toxic Product Available   Choose products made from safe, biodegradable materials. Select a water-based product over a solvent-based one for things like paint, glue, and shoe polish. Avoid aerosol products, which release fine particles of propellant every time they are used. When you inhale, these particles can lodge in your lungs.

Buy Only as Much as You Need   If you buy a gallon of paint when you need only a quart, you'll have to store the remainder. If it gets old and dried up, you may end up throwing it out with the trash.

Use as Directed   Use the recommended amount of a product, not more or less. Never mix chlorine-based cleaning products with those containing ammonia. If the material is flammable, extinguish nearby pilot lights. If it should be used in a well-ventilated area, open the doors and windows or turn on an exhaust fan. Take whatever precautions are necessary to protect yourself, by wearing goggles, gloves, or a face mask. Chemicals can enter your body by absorption through the skin or by inhalation of fumes and vapors.

Store Unused Portions Properly   Keep the product tightly sealed, upright, and in its original container. Make sure that toxic products are not stored near food and that children and animals cannot reach them. Keep flammable products away from a source of heat. Consider giving unused portions of substances like paint and paint strippers to friends and neighbors, a church, or a social service agency. Incidentally, if you have several partial cans of latex paint left over, you can mix the colors together. You'll usually end up with a beige or gray tint. It can be used as a primer coat or even as a final coat, if you like the color.

Recycle the Product If Possible   Some hazardous materials, such as paint thinners, motor oil, and car batteries, can be recycled. See page 30.

Dispose of Hazardous Wastes Properly   When you decide to get rid of unused portions of a hazardous material, check the label to see if it contains any directions for disposal. If so, follow them.

Some communities maintain a permanent hazardous waste collection center. Others set up temporary centers a few times a year to collect waste. If you're not certain what your community does, call the city public works department for information. Should the city have no provisions at all for dealing with hazardous waste, write letters to the mayor, other public officials, and the newspaper about the need for a collection center. And try not to buy any more hazardous products yourself.

Response: Use Environmentally Friendly Household Cleaners

Again, imagine being on a spaceship. Since you'd be living in close quarters, the fewer hazardous materials aboard, the better. The same thing applies to planet earth. Because our garbage will always be with us, we're all responsible for ensuring that as little of it as possible is hazardous to people, wildlife, or the environment.

Typical commercial household cleaners are toxic, corrosive, flammable, and/or irritant. Safe alternatives exist. Many natural food stores carry environmentally friendly, vegetable-based cleaning products. They can also be ordered from companies such as Seventh Generation (see page 21.) However, you can also make low-cost, nonhazardous cleaners yourself.

The five basic ingredients are baking soda (sodium bicarbonate), borax (sodium borate), pure soap (liquid or flakes), salt, and white vinegar. Ammonia, though biodegradable, is caustic and irritant; use it only when necessary. There are hundreds of recipes for these safe household cleaners, a few of which are given below. For those that you use often, you might want to make up a batch and keep it in a jar or pump-spray container.

All-Purpose Cleaner   Mix 1/2 cup baking soda, 1 cup vinegar, 2 cups warm water. (For appliances, walls, tile, etc.)

Carpet Cleaner, Deodorizer   Sprinkle baking soda liberally on a carpet, brush it in so it sinks to the bottom, then vacuum it up. Or mix 2 cups baking soda with 1/2 cup cornstarch and 4 crumbled bay leaves. Sprinkle on carpet and wait an hour or more before vacuuming. Fuller's earth, a clay powder, will absorb liquid spills on carpets and upholstery.

Copper and Brass Cleaner   Catsup removes tarnish from copper. You can also mix equal parts of vinegar, salt, and flour into a paste and coat the copper or brass object. Wait 10 minutes or more, then scrub, rinse, and buff with a soft, dry cloth.

Dishwashing Liquid or Powder   Use pure soap. For automatic dishwashers, use a commercial product low in phosphates or one labeled environmentally friendly.

Drain Cleaner   The best bet is not to let the drain become clogged in the first place. Never pour grease down the drain; pour it into an old can, let it harden in the freezer, and then discard it. Use a strainer on all drains.

It's also a good idea to pour boiling water down any drain once a week.

Should you have inherited a sluggish drain, try the following: Pour 1/2 cup baking soda, 1/2 cup vinegar, and 1/8 cup salt into the drain, let it settle, then pour in some boiling water. Repeat in 30 minutes if necessary.

Floor Cleaner   For linoleum or vinyl flooring, use a mild detergent or mix 1/2 cup vinegar with 1 gallon warm water. For wood floors, mix 1/4 cup oil soap with 1 gallon warm water.

Furniture Polish   Mix 1/4 cup mineral oil with 1/8 cup lemon juice. Apply with a soft cloth, then polish.

Garbage Can Cleaner   Mix 1/2 cup borax with 1 gallon warm water. Spray or sponge on, rinse, then let dry in sun.

Glass, Mirror, Window Cleaner   Mix 1/2 cup vinegar with 1 quart warm water in an empty pump-spray bottle. Apply to glass and rub dry with a rag or newspaper. For heavily soiled or greasy glass, mix 1 cup ammonia and 3 cups water. Wear rubber gloves and ventilate the room.

Mildew Cleaner   Scrub mildew spots with mixture of 1 cup borax and 1 cup vinegar. In an automatic washer, use 1/2 cup soap and 1/2 cup baking soda to clean a shower curtain, adding 1 cup vinegar to the rinse cycle.

Oven Cleaner   To catch spills, line the bottom of the oven with aluminum foil. To clean a dirty oven, mix 1/4 cup baking soda with 1 quart warm water. Apply, wait 20 minutes, then clean. For stubborn stains, mix 1/2 cup ammonia, 1/4 cup baking soda, and 1/2 cup vinegar into a paste. Leave it on the stain for 20 minutes, then scrub with a sponge or very fine steel wool.

Sink, Tub, and Tile Cleaner   Use baking soda and water. Or put a plug in the sink, pour in 1/2 cup vinegar, sprinkle on baking soda, and then scrub.

Toilet Cleaner   Sprinkle baking soda on your toilet brush and swish it around the toilet bowl. Or pour 1/2 cup borax into the bowl and let it stand for an hour before scrubbing and flushing. You can remove stubborn rings with a pumice stone.

Response: Know the Hazardous Materials in Your Garage and Workshop

A number of hazardous products are found in the typical garage and workshop. Avoid buying them when possible, and learn how to dispose of those you cannot avoid.

Antifreeze   Contains ethylene glycol, which is extremely toxic and should not be poured down a storm sewer or on the ground. Find a service station that cleans and reuses antifreeze, or else dispose of used antifreeze as hazardous waste.

Car Batteries   Batteries contain lead and sulfuric acid, so don't dispose of them with your regular trash. Take your old battery to a garage for recycling or to a hazardous waste collection center.

Motor Oil   If you change your own oil, do not dump waste oil down the sewer or on the ground; it can contaminate soil and water supplies. Furthermore, the used oil is a valuable resource that can be re-refined and reused. Take your waste oil in a sealed container to a garage that will recycle it or to a hazardous waste collection center. Buy re-refined motor oil.

Transmission and Brake Fluids   Recycle at participating service stations or dispose of as hazardous waste.

Paints and Stains   Whenever possible, use water-based, latex paint instead of oil-based paint because latex paint contains far fewer volatile organic compounds. Be sure, however, that the latex paint does not contain mercury preservatives, which can cause mercury poisoning. Use water-based stains and varnishes.

Paint Stripper   Highly toxic. Use a scraper and heat gun instead.

Paint Thinner, Turpentine   If you do use an oil-based paint, you can recycle the thinner yourself. After you've cleaned your brush, store the used thinner in a jar. After the paint particles have settled to the bottom, pour the clear liquid from the top into another jar. Seal, label, and store until you need it.

Wood Finishes   Linseed oil, shellac, and tung oil are all derived from natural sources and are safe to use. They do need to be diluted with an alcohol solvent or turpentine.

Response: Substitute Natural for Chemical Biocides

The products we use to try to rid our houses and yards of pests comprise a significant source of hazardous materials. Known collectively as **biocides,** they include insecticides, pesticides, rodenticides, fungicides, and herbicides. They are meant to kill living organisms; in fact, the suffix "cide" comes from the Latin verb meaning "to kill."

There are a number of very good reasons why we should use as few chemical biocides as possible:

- They harm creatures other than those whom they're intended to kill. Once used, a biocide settles into the soil, where it remains or is washed into a body of water. In either case, it is absorbed by organisms living in the soil or the mud. Through a process known as **biological magnification,** the biocide accumulates and is concentrated at progressively higher levels in the food chain. Predators accumulate larger amounts than their prey, and the effect on birds and fish may be lethal.

  Some commonly used chemical biocides are suspected of causing cancer and birth defects and damaging the kidneys, liver, nervous systems, and immune systems in humans. Even the chemicals used in some flea collars have been found to cause permanent nerve damage, cancer, and birth defects in cats and dogs.

- Biocides end up in our food supply. More than 100 different pesticides have been detected on commonly eaten fruits and vegetables.

- Biocides may exacerbate the problem their use was designed to eradicate. By altering the natural processes that determine which insects in a population will survive, biocides spur the development of resistant species. If all but 5% of the mosquito population in an area is killed by an insecticide, the ones that survive are the most resistant individuals, and they are the ones that will produce the succeeding generations. More than 400 insect and mite species are known to be resistant to pesticides, and some "super weeds" are totally resistant to certain herbicides.

- Biocides have also created further problems by destroying beneficial insects, the natural enemies of the intended target, leaving the pest to breed unchallenged. Thus, spraying an infested crop might kill 90% of the pests—but it also kills the insects that eat the pests. With their food abundant and their predators rare, the remaining pests recover faster than their enemies, whose prey is now scarce and harder to locate.

- Biocides are overused. American households annually use over 60 million pounds (27 million kg) of toxic chemical biocides; agricultural uses account for another 460 million pounds (209 million kg).

- The entire process of insecticide development may be self-defeating. Despite the billions of pounds (and dollars) used to attempt to eradicate pests, crop loss to insect and weed pests has actually grown. According to Department of Agriculture figures, 32% of crops were lost to pests in 1945; forty years later, such losses had increased to 37%.

The solution for us as individuals is to avoid using commercial chemical biocides and to try alternative ways to get rid of pests:

- Use your own organic pesticides. Derived from plants, they don't persist in the environment as long as do chemical pesticides. Thus cockroaches can be controlled by keeping your kitchen as clean as possible and sprinkling borax in cracks and crevices around baseboards, appliances, and ducts. Don't put it where children and pets can get at it, however. Get rid of aphids by spraying plants with a mixture of pure soap dissolved in hot water. If you have a pet that is bothered by fleas and ticks, make the following rinse: Add 2 tablespoons of rosemary to 2 cups of boiling water. Let it steep for 20 minutes, strain it, and allow it to cool. Spray or sponge the mixture on your pet and allow it to air dry. Some people have reported success with feeding their pets either brewer's yeast or a liver-flavored vitamin B complex supplement, which makes the animal unappealing to fleas.

- Purchase environmentally friendly alternatives. A number of companies market natural pest control products, including insect repellents, insect traps, and flea and tick powders and sprays. Check your local health food or ecology store. Citrus rind oil concentrate is available at some pet stores. It can be used as a flea dip for dogs and cats.

- Buy organically grown fruits, vegetables, and grains. These have been grown without the use of chemical biocides. You'll not only be supporting farmers who are committed to protecting the environment, you'll also be ingesting fewer chemical residues yourself.

CHALLENGE: DESTRUCTION OF THE TROPICAL RAIN FORESTS

Have you ever enjoyed being in a forest and breathing the cool, oxygenated air? The wonder of the world's trees is that they extract more carbon from atmospheric carbon dioxide and release more oxygen back into the atmosphere than any other type of vegetation.

When the peoples of Europe and America were busy destroying their forests, not much was made of it because in the nineteenth century little was known about the enormous importance of the oxygen that forests produce. Nowadays it is well known that the carbon-to-oxygen process is important. Without it, carbon would build up in the atmosphere, speeding up the greenhouse effect (see pages 58–60). Nowhere is this process more important than in the tropics.

In recent years, newspapers and magazines have devoted considerable space to the topic of the destruction of tropical rain forests. The Amazon Basin has the greatest forests in the world. The basin is as large as the United States. It and the large forests of Africa and Southeast Asia are the last remaining tropical forests in the world. Because the warm, moist environment is more conducive to photosynthesis and therefore more oxygen production, tropical rain forests are environmentally more important than the northern forests.

The Amazon Basin lies chiefly in Brazil. Brazilians have been anxious to develop their country so that they can increase their standard of living and international prestige. Brazilian planners have looked inland for the materials of growth and development, just as resourceful Americans and Canadians of the eighteenth and nineteenth centuries spread across North America looking for wealth. After building a new capital city, Brasilia, 350 miles (560 km) from the coast, Brazilian leaders began opening the Amazon region for cattle grazing, farming, and mineral exploitation. Many poor Brazilians in the dry lands of the Northeast, faced with a life of marginal subsistence farming, moved to the interior forest. They leveled the forest by burning it. Brazilian scientists estimate that there were 170,000 separate fires in the Amazon in 1987 alone. Likened by some to an environmental holocaust, the fires generate hundreds of millions of tons of gases that contribute to global warming and depletion of the earth's protective ozone layer. At this very moment, in this one second, a stand of magnificent trees the size of a football field has been destroyed by fire.

The forests are one component in an intricate ecosystem that has developed over millennia. Destroying forests undermines a major part of the biological diversity of the planet. Once the forests are removed, many plant and animal habitats can no longer exist. Of the millions of plant and animal species of earth, about 40% are native to the tropical rain forest. It has been estimated that from one to twenty-four species in the forest become extinct every day.

After the forests are burned, the new farmers and ranchers, unused to caring for the soil of the rain forest, fail to employ conservation practices. Topsoil, which is so susceptible to erosion in warm, wet regions, is easily washed away or depleted. In fact, in Brazil, farmers are finding that agricultural yields and the quality of cattle decrease rapidly after the first year of production.

In other parts of the world, especially in the developing countries of Southeast Asia (Indonesia, Malaysia, the Philippines) and in Africa, major hydroelectric projects and logging are going ahead with little concern for the global effects of depleted forests. Only recently has it been discovered that when forests are cleared on a grand scale, more sunlight is reflected off of the earth's surface. Wind currents and rainfall patterns are disrupted. There is now evidence that forest clearance leads to desertification in some areas.

In addition to the perceived need for forest exploitation in developing countries, there is at the same time an enormous demand in the developed world of North America and Europe for tropical hardwoods (mahogany, teak, ebony, and rosewood) and for imported beef derived from the cattle that now graze in what was once a tropical rain forest. In addition, the pharmaceutical industry has long depended on tropical plants for making invaluable medicines. Future discoveries of new medicines will be limited by forest depletion and species extinction.

The destruction of the rain forests is a tragedy that yields no long-term benefits. The world is fast approaching the end of a period in which resources were cheap, readily available, and lavishly used. Too much is at stake for us to sit idly by while the Amazon burns.

Response: Help Save the Tropical Rain Forests

To be honest, saving the tropical rain forests from destruction is largely up to the countries themselves and to lending institutions that help fund their development projects. There is relatively little the individual American can do—but there is still something. You might consider the following list of do's and don'ts:

- *Don't* patronize fast-food restaurants that purchase beef from cattle grazing on deforested rain forest land. American fast-food chains have imported large amounts of beef from Central and South America because it is less expensive than domestically produced beef. But it takes 55 square feet of cleared rain forest land to produce enough hamburger to make a quarter-pounder that will then cost you a nickel or dime less than it would have otherwise. It's not a bargain.

- *Do* buy Brazil nuts, cashews, and other tropical nuts that require an intact, living forest for their continued production. Harvesting the nuts enables people who live in the tropical forests to earn a living, and it gives them an incentive to preserve the forests. Rainforest Crunch, a cashew and Brazil nut brittle, is marketed by Community Products, Inc., which donates 40% of the profits to organizations helping to protect the Amazon rain forest.

- *Don't* buy tropical hardwoods or products made from them. These woods include ebony, mahogany, rosewood, and teak, and they are used for a variety of products: bowls and carvings, furniture and veneer, fiberboard and plywood, and even wooden toilet seats. Many large timber companies, including Weyerhauser and Georgia Pacific, import tropical rain forest timber; this spurs deforestation. Purchase products made from domestic woods instead, such as ash, birch, cherry, maple, oak, and pine.

- *Do* support groups that are involved in saving the rain forests. These include, among others, Friends of the Earth, Lighthawk, Nature Conservancy, Probe International, Rainforest Action Network, Rainforest Alliance, World Resources Institute, and the World Wildlife Fund. The latter organization has a unique "Guardian of the Rainforest" campaign. For $25 you can "buy" an acre of Amazon rain forest and save it from destruction. The money will be used to pay wardens to patrol sensitive areas and to help local people develop sustainable uses of the forest, such as rubber tapping and harvesting fruits and nuts.

THE WATER

CHALLENGE: SAVING WATER

Comparing the earth to a spaceship is especially appropriate in the case of water. All of the water needed for a journey through space would have to be loaded on board at the beginning of the trip. There would be no way to manufacture additional water once the ship was in orbit. The water would have to be used over and over again.

In exactly the same way, the supply of water on earth is constant, and it's over 4 billion years old. The system by which water is continuously circulated through the biosphere is called the **hydrologic cycle.** Evaporation and transpiration (the emission of water vapor from plants) are the mechanisms that redistribute water. Water vapor collects in clouds, condenses, and then falls again to earth. There it is reevaporated and retranspired, only to fall once more as precipitation.

If the supply of water is constant, if it will always be there, you might ask why we need to worry about conserving it. There are two good reasons. First, while we're not about to run out of water, we're already running out of inexpensive water. Think of what had to happen before you could turn on the faucet and get water. People had to locate a source of water; they had to build aqueducts, canals, water tunnels, and pipes to carry it, machines to pump it, and plants to treat it. To carry away used water, they've had to build drains, sewers, and more plants to treat the water before it's discharged into a stream, lake, or ocean.

Providing water is costly, and the cost isn't just monetary. For example, the once majestic Colorado River, the only significant source of surface water in the southwestern United States, has been so transformed by dams, canals, pipelines, and reservoirs that it is now little more than a vast, controlled plumbing system. By the time the 1,400-mile-long (2,240-km) river reaches Mexico, it isn't much more than a creek. There a final dam, the Morelos, diverts for irrigation what little water remains, so that by the time the river reaches the sea, it is dry. This Colorado River is a far cry from that seen by the Spanish explorer Hernando de Alarcon, who in 1540 journeyed up the river from the Gulf of California and described it as, "A very mighty river, which ran so great a fury of a storm that we could hardly rail against it."

The fate of the Colorado River implies the second good reason for conserving water. Agriculture, industry, urbanization, and a growing population are placing increasing demands on water supplies. As pressure on the water supply increases, there *will* be regional water shortages.

Thirty dams tame the Colorado and its tributaries, and two massive reservoirs (Lake Mead and Lake Powell) store its water. Aqueducts and irrigation canals siphon off its water for use in seven western states and northern Mexico. Los Angeles, Denver, and hundreds of other cities couldn't exist as they are today without the water the Colorado provides, and the fruits and vegetables raised in the Central and Imperial Valleys would die.

Legally, all the water in the Colorado River is spoken for. California receives 4.4 million acre-feet per year, Arizona 2.8 million acre-feet, and so on. (One acre-foot is the amount of water that would cover an acre of ground with a foot of water—about 326,000 gallons.) Indeed, yearly allotments now stand at 16.5 million acre-feet, in spite of the fact that the Colorado rarely carries more than 14 million acre-feet! Shortages haven't occurred yet simply because not all the states are using the full share to which they are entitled. But demands are expected to increase to that point in the very near future, perhaps by the end of this decade.

Another area where the demand for water is beginning to exceed the supply is the High Plains region, which stretches from South Dakota to Texas. Agriculture there depends on drawing irrigation water from a vast underground formation called the Ogallala Aquifer. (An aquifer is zone of water-saturated sands and gravels beneath the earth's surface.) The Ogallala Aquifer is the country's largest underground water supply, spreading under 20 million acres in eight states. It supports nearly half the country's cattle industry, a fourth of its cotton crop, and a great deal of its corn and wheat. But the 150,000 wells that now puncture the aquifer cause the water table to fall from 2 to 5 feet (.6 to 1.5 m) each year—a rate that is far greater than the rate at which the aquifer can be replenished by nature. Hydrologists expect that as much as 40% of the irrigated acreage will be lost in the next twenty years, causing economic hardship in the region and food shortages in the country.

Response: Flush and Brush with Care

The average American uses about 100 gallons of water a day. Most of that water is used in the bathroom, and much of it is used needlessly. There are some very simple things you can do to save almost half the water you use. Several of them take just a few minutes to accomplish.

Save Water When Flushing   Of the 100 or so gallons of water you consume a day, about 40 are used to flush the toilet. The typical toilet uses from 5 to 7 gallons—which is why flushing uses more water than anything else in your apartment or house. Here are some suggestions for conserving water when flushing:

- The easiest way to cut down on the amount of water is to simply flush less often, but most of us don't want to do that. Another way is to fill a large plastic container with water, add an inch or two of sand or pebbles to weigh it down, put on the cap, and then insert it in the toilet tank. Depending on the configuration of your tank, you might want to use several smaller containers instead of one large one. Experiment to see what works best and to determine how well the toilet flushes with the bottle(s) in place. You can put in a brick, but be sure to wrap it in plastic so that, if it disintegrates, particles won't clog the toilet.

     How much water will you save? The amount you displace. A one gallon bottle will save one gallon of water each time the toilet is flushed.

- An alternative to the plastic bottle is a displacement bag, a plastic pouch that you fill with water and hang inside the tank. Plumbing and hardware stores have them. Also available at these stores are toilet dams, plastic barriers that you wedge inside the tank on either side of the flush valve. The dam prevents the water in that section of the tank from running out when you flush. A typical dam saves two gallons of water.

- More expensive ($20–30), but also more effective at saving water, is a device that provides a short flush (1.6 gallons) for liquid waste and paper and a fuller flush (3.5 gallons) for solid waste. The kit can be installed without special tools.

- An adjustable float assembly lets you reduce the water level in the tank. If you do adjust the float, test to make sure that wastes will still be carried away with a single flush.

- The single best solution is the state-of-the-art ultra-low-flush toilet. These use only 1 to 1.6 gallons per flush, and if you or your landlord is putting in a new toilet, this is what should be installed. It has been estimated that if the entire southern coast of California installed low-flush toilets, 5,000 acre-feet of water (1.6 *billion* gallons) would be saved a year.

- Finally, check for leaks. A leaky toilet could be wasting 50 gallons or more per day. Add a few drops of food coloring to the tank, then wait 15 minutes. If the tank is leaking, the water in the bowl will change color.

Turn Off the Tap

- If you leave the tap on while you brush your teeth, 2 to 5 gallons of water are likely to run down the drain. The solution: Turn on the water, wet your toothbrush, then turn off the tap until it's time to rinse. You'll use one cup of water instead of gallons.

- Leaving the faucet running while you shave will use anywhere from 5 to 20 gallons of water. The solution: Fill the basin and turn off the tap until it's time to rinse your face.

- Turn off the tap while you wash your hands, and you'll save another 2 or 3 gallons.

Buy Low-Flow Faucet Aerators   Low-flow aerators are simple devices that mix air and water as it comes out of the tap. They reduce the flow by about half, but you won't notice the difference because they make a small stream of water seem plentiful. Don't assume that your sinks have them just because there is a screen aerator on the faucet. If water comes out of the tap at 3 gallons per minute or more, you need a low-flow model. These cost less than $5, will fit both bathroom and kitchen sinks, and are easy to install. You just unscrew the old aerator, if there is one, and screw on the low-flow model.

Response: Save Water While You Bathe

The opportunities to save water while bathing are legion. The amount you save depends on how often and how long you bathe and also on the rate at which water comes out of your showerhead. Here are some tips:

- Bathe less often. Most people don't really need to bathe every day—and wouldn't if water were priced at $1 per gallon or if they had to haul it from outside themselves.

- A bath or shower—which uses more? The average bath consumes about 40 gallons of water. If you take a 2- or 3-minute shower, you'll use less than that, but a 10-minute or longer shower will use more. If you take a bath, keep the water level low. If you shower, keep it short. A 10-minute shower uses twice as much water as a 5-minute shower. At 5 gallons per minute (gpm), that's 25 gallons saved.

- With a one-time investment of $10–15 in a low-flow showerhead, you can save half or even more of the water you use to bathe and reduce your cost of heating hot water at the same time. Low-flow showerheads reduce water flow to 2 or 3 gallons of water a minute while still delivering water with invigorating force. They fit most showers, come in both aerated and nonaerated models, and don't require special tools for installation.

    How do you know if you need a low-flow showerhead? Hold a half-gallon container directly under your shower while it's running, making sure to collect all the water. If it fills in less than 10 seconds, purchase a low-flow model.

    How much water will you save? If you have a shower that delivers 5 gpm, and you take a 5-minute shower, you'll save 15 gallons a day by using a 2 gpm low-flow showerhead. That's 5,475 gallons per year. If you're used to taking a 10-minute shower and the tap delivers 8 gpm, you'll save 70 gallons a day with a low-flow model—or 25,550 gallons per year! Who says one person can't make a difference?

Response: Other Ways to Save Water

Washing Dishes   If you let the water run while you wash dishes by hand, you'll use 20 to 30 gallons of water. But if you fill the sink or a washtub with soapy water and turn on the tap only to rinse the dishes, you'll cut your water usage to 10 gallons.

Automatic dishwashers can be water-savers. They typically use 20–25 gallons of water. Run the machine only when it's full, and run it on the "short" or "light" cycle. That shouldn't consume more than 10 gallons of water.

Washing Clothes   A top-loading automatic washer run at full cycle uses 50 to 60 gallons of water. Use it only when you have a full load, or else adjust the water-level setting. The short or gentle cycle is perfectly satisfactory for typically soiled clothes. It uses only 25 to 30 gallons of water, a savings of about half. You'll also be reducing the cost of heating so much hot water.

Washing the Car   The most water-efficient way to wash your car is to fill a bucket with soapy water and use a hose only to rinse.

If you use a hose, make sure it has a shut-off nozzle. If you simply put the hose on the ground and let it run while you wash the car, 100 gallons or more are likely to be wasted.

If you go to an automatic car wash, patronize one that recycles the water.

Watering the Lawn, Trees, and Shrubs

- It's best to water outside before 8 A.M. or after 6 P.M. to prevent the sun from evaporating much of the water before it can soak into the ground.

- In hot weather, let the grass grow a little taller before you mow it—or set the lawn mower blade a notch higher. Longer grass means less evaporation, so you can save hundreds of gallons of water a month.

- Position your sprinklers so as to avoid watering the sidewalk or driveway.

- To retain as much moisture as possible, put mulch around plants and dig a shallow basin around trees and shrubs to prevent runoff.

CHALLENGE: WATER POLLUTION

We said earlier that the supply of water is constant, and that the system by which it continuously circulates through the biosphere is called the hydrologic cycle. In that cycle, water may change form and composition, but, under natural environmental circumstances, it is purified in the recycling process. When water composition has been so modified that either it cannot be used for a specific purpose or it is less suitable for that use than it was in its natural state, the water is said to be **polluted.**

Pollution is caused by discharging into water substances that cause unfavorable changes in its chemical or physical nature or in the quantity and quality of the organisms living in the water. Pollution is a relative term. Water that is not suitable for drinking may be completely satisfactory for cleaning streets.

People are not the only cause of water pollution. Decayed leaves, animal wastes, and other natural phenomena may affect water quality. There are natural processes, however, to take care of such pollution. Organisms in water are able to degrade, assimilate, and disperse such substances in the amounts in which they naturally occur.

What is happening now is that the quantities of wastes discharged by people often exceed the ability of a given body of water to purify itself. In addition, we are introducing pollutants, such as metals or inorganic substances, that take a very long time to break down or that cannot be broken down at all by natural mechanisms.

The four main contributors to water pollution are agriculture, industry, mining, municipalities and residences.

Agriculture

The kinds of pollutants associated with agriculture are biocides (see page 31), fertilizers, and animal wastes. Runoff from farms and feedlots carries these contaminants into underground and surface waters. Fertilizers are responsible for depositing excess nutrients (nitrates and phosphates) in water bodies. There they hasten the process of **eutrophication.** Algae and other plants are stimulated to grow abundantly. When they die, the level of dissolved oxygen in the water decreases. Fish and plants that cannot tolerate the poorly oxygenated water are eliminated. Symptoms of a eutrophic lake are prolific weed growth, large masses of algae, fish kills, and water that has a foul taste and odor. About one-third of the medium- and large-size lakes in the United States have been affected by accelerated eutrophication.

Industry

Many industries dump organic and inorganic wastes into bodies of water. These may be acids, highly toxic minerals, or, in the case of petroleum refineries, toxic organic chemicals. The nuclear power industry has caused some water pollution when radioactive material has seeped from tanks containing nuclear wastes. Acid rain (see page 53), a by-product of emissions from factories, power plants, and motor vehicles, has affected the water quality of thousands of lakes and streams.

Mining

Surface mining for coal, iron, copper, gold, and other substances contributes to contamination of the water supply through the wastes it generates. Rainwater reacts with the wastes, and dissolved minerals seep into nearby water bodies. In addition to altering the quality of the water, contaminants affect plant and animal life. Each year, for example, thousands of animals and migratory birds die in such western states as Arizona, Nevada, and California after drinking cyanide-laced waters at gold mines.

Municipalities and Residences

A variety of pollutants comes from the activities associated with towns and cities. The use of detergents has increased the phosphorus content of rivers, and salt (used for de-icing roads) increases the chloride content of runoff. Water runoff from urban areas contains contaminants from garbage, insecticides, animal droppings, litter, vehicle drippings, and the like. Because the sources of pollution are so varied, the water supply in any single area is often affected by diverse pollutants, which complicates the problem of controlling water quality.

Contaminated drinking-water wells have been found in more than half of the states. Thousands of wells that tap aquifers have been closed in such states as New York, New Jersey, Massachusetts, and California because of chemical contamination. Chemicals have reached groundwater by seeping into aquifers from landfills, fuel storage tanks, sewers, and runoff from paved surfaces in urban areas. The pollution of aquifers is particularly troublesome because we depend on them for about half of our drinking water.

Sewage can be a major water pollutant, depending on how well it is treated with chemical disinfectants or filters before being discharged. This is not simply an environmental concern; it directly affects human health. Raw, untreated human waste contains viruses responsible for dysentery, hepatitis, spinal meningitis, and other diseases. Only half of the American population lives in communities with sewage-treatment plans that meet the minimum goals set by the federal Clean Water Act. Aged sewer systems in more than 1,100 cities still discharge poorly treated sewage into streams, lakes, and oceans.

Response: Launder Your Clothes Responsibly

You may be contributing to the demise of streams and lakes without even realizing it.

- Many detergents contain excessive amounts of phosphates. When the sewage system deposits the waste water in bodies of water, the phosphates stimulate the growth of algae and other plants. The end result is a decrease in the oxygen content of the water, fish kills, and, in extreme cases, a river or lake devoid of life. Some states have banned the use of phosphates in detergents, and others limit the amount of phosphorus a detergent may contain.

- Chlorine bleaches increase the chlorine content of water bodies, to the detriment of plant and animal life.

- Some pretreatment stain removers contain toxic ingredients.

You can get your clothes clean without harming the environment:

- The side of your box of detergent contains information about the amount of phosphorus in the form of phosphates that it contains. Multiply the percentage of phosphorus by three to get the phosphate content. If the box says 5% phosphorus, it contains 15% phosphates.

- If the amount of phosphorus is over 2%, switch to a different brand. Some contain no phosphorus at all, others are as low as 0.5%. Most liquid detergents are phosphate-free.

- An alternative is to use pure laundry soap (powder or flakes) instead of detergents. If your water is hard, add borax along with the laundry soap.

- If you use a commercial bleach, buy one without chlorine. In most cases, you can whiten whites and brighten colors by adding 1/2 cup borax to the washload.

- Here is a recipe for a pretreatment spray that is a winner on 3 counts. It is not harmful to the environment; it does not come in an aerosol can; and it is inexpensive. Ingredients: 1/4 cup ammonia, 1/4 cup white vinegar, 1/8 cup baking soda, 1 tablespoon liquid soap, 1 quart water. Mix, label, and store in a pump-spray container.

Response: Three Other Ways to Keep the Water Clean

Using environmentally friendly cleaners to launder your clothes is one way you can help keep our water supply unpolluted. Here are some other suggestions:

Avoid Excessive Use of Lawn Fertilizers   Americans tend to like velvety-looking, dark green grass. In pursuit of that goal they annually apply millions of pounds of fertilizer to their lawns—far more than is needed. Wind and rain sweep the excess nutrients into streams and lakes.

Most commercial fertilizers contain a combination of nitrogen, phosphorus, and potassium. Both nitrogen and phosphorus harm bodies of water. Phosphorus contributes to algae bloom and eutrophication (see page 43); nitrogen deprives the water of oxygen and results in fish kills.

Before applying any fertilizer, have your soil tested to see what it actually needs. Garden stores and county extension agents perform soil analyses and tell you what you should apply, how much, and when. Apply only the amount indicated. It may well be that you won't need to buy chemical fertilizers. Nonchemical fertilizers include manure, bonemeal, and compost.

Purchase Unbleached Paper Products   Whenever you have a choice, buy unbleached, undyed goods such as paper napkins, towels, and coffee filters. Paper mills contribute significantly to the pollution of rivers when they use chlorine to bleach paper pulp. The manufacturing process generates dioxin, a highly toxic contaminant, and other chlorinated hydrocarbons. The fewer bleached and dyed products that are manufactured and disposed of, the fewer pollutants will end up in the water supply.

Dispose of Hazardous Wastes Properly   This topic is discussed on pages 26–27. Remember that anything you pour on the ground, into a storm sewer, into your sink, or down the toilet will eventually end up in the water supply. Remember, too, that many communities dump their raw, untreated sewage directly into rivers, lakes, and oceans. Use as few hazardous products as possible, and, when you need to dispose of unused portions, take them to a collection site.

Response: Help Halt Marine Pollution

About half of the population of North America lives within 100 miles (160 km) of the Atlantic, Gulf, or Pacific coast, and the percentage is growing. The presence of so many people is putting an increasing strain on coastal areas.

One of the primary polluters of coastal waters is human sewage. Many communities drain raw, untreated sewage into harbors and ocean water through outfall pipes. Others annually dump millions of gallons of treated sewage, called sludge, into the ocean. Beaches are periodically closed when the water is deemed to be unsafe for swimming.

Though not harmful to humans, plastic is also a pollutant of coastal waters. Some of the plastic has been disposed of on land and then washed into coastal waters through streams and sewer systems. Some has been blown into the water from nearby landfills, and yet other plastic waste is simply litter that people have carelessly left on beaches.

Every year more than a million birds and thousands of fish, seals, otters, and turtles die because of plastic. Lightweight and buoyant, plastic bags and polystyrene containers float on the water, where birds, fish, and other sea creatures mistake them for food. These animals die from choking and malnutrition. They also get ensnared in transparent plastic fishing lines, nets, and six-pack rings and die from starvation.

You can do your part to keep coastal areas clean:

- Don't make the sewage system absorb wastes it can't handle. Don't pour any hazardous wastes on the ground, into a storm sewer, or into a sink or toilet.

- Use as little plastic as possible. Don't leave plastic bags or foam cups, plates, and fast-food containers on the beach. Cut apart six-pack rings before disposing of them so that they can't ensnare birds and fish.

- If you see litter on a beach, pick it up. Leave the area as clean or cleaner than it was before you came.

THE AIR

CHALLENGE: SMOG

If you live or have traveled in the United States, the chances are pretty good that you've encountered smog on at least one occasion. More than one-third of the population lives in metropolitan areas that have smog one or more times during the year.

Smog is a hazy mixture of gases and solid particles. While it may contain as many as one hundred different chemical compounds, three ingredients are crucial to its formation: nitrogen oxides, hydrocarbons, and sunlight. Smog is created when oxides of nitrogen react with the oxygen present in water vapor in the air to form nitrogen dioxide. In the presence of sunlight, nitrogen dioxide reacts with hydrocarbons to form new compounds, such as ozone. Ozone is *the* major component of smog. As long as sunlight continues to bake the mixture of hydrocarbons and nitrous oxides, smog continues to form. That's why ozone levels tend to be worse in the summer months than during the rest of the year.

Although nitrogen oxides and hydrocarbons are generated by a variety of sources, the chief culprits are motor vehicles. They are responsible for about 50% of smog. Engine exhaust contains both hydrocarbons and nitrogen oxides. In addition, hydrocarbons come from gasoline vapor that escapes when you refuel your car.

Other sources of smog are drycleaning establishments (the solvents emit fumes), furniture makers and paint shops (which use hydrocarbon solvents, paints, and varnishes), and bakeries (sunlight changes yeast by-products into ozone). These establishments, as well as consumer products such as aerosol hairsprays and deodorants, account for another 40% of smog formation. Power plants and industrial facilities typically contribute 15% of the ingredients found in smog.

Although all of America's largest cities have too much smog, four regions that are characterized by high ozone levels are California, the Texas Gulf Coast, the Great Lakes area, and much of the Northeast. A number of factors determine whether an area will have smog. Clearly, the size and density of the population, the amount of traffic, and the number of industries are crucial since they all help determine the kinds and amounts of substances discharged into the air.

Unusual weather can also be a factor if it alters the normal pattern of pollutant dispersal. A temperature inversion will magnify the effects of dirty air. This occurs when a stationary layer of warm air over a region prevents the normal rising and cooling of air from below. The air becomes stagnant. As pollutants accumulate in the lowest layer instead of being blown away, the air becomes more and more contaminated. Temperature inversions occur often in the Los Angeles basin in the fall and in Denver in the winter.

Also, the air pollutants generated in one place may have their most serious effect in areas hundreds of miles away. Thus, the worst effects of the smog that originates in New York City are felt in Connecticut and parts of Massachusetts. The chemical reaction that produces ozone takes a few hours, and by that time air currents have carried the pollutants away from the city.

If you've experienced a really smoggy day, you know that smog makes your eyes sting and your throat feel raw. But you may not be aware of its more harmful effects. Ozone reduces a person's lung capacity, scars lung tissue, and lowers the lungs' resistance to infection. Ozone is especially hazardous to children (because of their small breathing passages), to the elderly, and to those who already suffer from asthma, emphysema, and other respiratory diseases.

Smog is also a significant cause of damage to crops and trees. Because ozone damages plant cell membranes, crops grown under smoggy conditions mature more slowly and yield less than normal. The Environmental Protection Agency has estimated that ozone pollution reduces crop yields by as much as $3 billion per year.

Response: Control Harmful Emissions

If You Drive   Motor vehicles are the single most important contributor to smog formation, accounting for emissions of both hydrocarbons and nitrogen oxides. You can help prevent smog by using your car as little and as efficiently as possible. We offer some suggestions for reducing gas consumption here; others appear on pages 55–56.

- Think before you drive. Can you walk, bike, or take mass transit where you want to go? Can you combine many short trips into a single, longer one? Can you carpool?

- Have the car tuned every 5,000 to 10,000 miles (8,000 to 16,000 km), and make sure its antismog equipment is functioning properly. If you have a newer car, it probably has a catalytic converter, a cylinder full of metal-coated ceramic pellets through which exhaust gas from the engine is funneled. Its function is to remove hydrocarbons and, to a lesser extent, nitrogen oxides from the exhaust.

- Make sure your tires are properly inflated. Underinflated tires increase resistance and thus waste gas.

- Don't let the car idle unnecessarily when you're starting up, waiting at a train crossing, or talking to a friend.

- Try to maintain a steady speed on the highway, preferably 55 miles per hour. You'll use considerably less gas than at higher speeds.

- If you can find a service station that sells it, try using methanol or gasohol in your car, especially if you live in a region that has smog. They produce cleaner exhaust gases than does gasoline.

- If you have an old car that uses leaded gas, you're spewing huge amounts of nitrogen oxides (and lead particulates) into the atmosphere. Ask your service station whether the car can run on unleaded gas.

- Many states now require that gas pumps be fitted with vapor-recovery nozzles. They make a tight seal with the car's filling pipe and ensure that fuel vapors (which contain hydrocarbons) are returned to the underground tanks at the gas station. Don't pull back the rubber seal.

- Don't "top off" the gas tank. Stop when the nozzle clicks off. Topping off not only forces gas fumes into the atmosphere, it's also likely to result in spilling some gas. The fumes and spilled gas help create smog.

Other Things to Do   We asked that you think twice before driving anywhere. Also think twice before using any of the following, because all of them emit chemicals that combine to form smog:

- A gas-powered lawn mower. (Electric or hand mowers are good alternatives.)

■   An oil-based paint. Use latex instead. Oil-based (alkyd) paints contain solvents that evaporate as the paint dries; in the air, they react with other gases and sunlight to form ozone.

■   Lighter fluid to start your barbecue grill. (Try an electric lighter.)

■   Any aerosol spray, whether it's hairspray, deodorant, oven cleaner, or paint.

CHALLENGE: ACID RAIN

Acid rain is the term generally used for pollutants that are created by burning fossil fuels and that then change chemically as they are transported through the atmosphere and fall back to earth as acidic rain, snow, fog, or dust. (Actually, "acid precipitation" is a more precise description.) The pollutants are chiefly oxides of sulfur and nitrogen, and they come primarily from coal- and oil-burning power plants and industries and from automotive exhausts. When sulfur dioxide is absorbed into water vapor in the atmosphere, it becomes sulfuric acid. Sulfur dioxide contributes about two-thirds of the acids in the rain; about one-third come from nitrogen oxides, transformed into nitric acid in the atmosphere.

Once the pollutants are airborne, winds can carry them hundreds of miles, depositing them far from their source. Much of the acid rain that falls on the eastern seaboard and eastern Canada originates in ten states in the central and upper Midwest. Similarly, airborne pollutants from Great Britain, France, and Germany cause acidification problems in Scandinavia.

Acid rain has three kinds of effects: terrestrial, aquatic, and material. When the acids are washed out of the air by rain, snow, or fog, they change the pH (potential of hydrogen) factor of both soil and water, setting off a chain of chemical and biological reactions. The pH factor measures the acidity/alkalinity of a substance on a scale of 0 to 14. It is important to note that the pH scale is logarithmic, which means every step on the scale represents a factor of ten. Thus 4.0 is ten times more acidic than 5.0 and one hundred times more acidic than 6.0. The average pH of normal rainfall is 5.6, but acid rainfalls with a pH of 1.5—far more acidic than vinegar or lemon juice—have been recorded.

Acid deposition harms soils and vegetation in part by coating the ground with particles of aluminum and toxic heavy metals such as cadmium and lead. It kills microorganisms in the soil that break down organic matter and recycle nutrients through the ecosystem. Significant forest damage has occurred in the eastern United States, northern and western Europe, Russia, and China.

The aquatic effects of acid precipitation are manifold. The problem is that the acidity of a lake or stream need not increase much before it begins to interfere with the early reproductive stages of fish. Also, the food chain is disrupted as acidification kills the plants and insects upon which fish feed. Acid rains have been linked to the disappearance of fish in thousands of lakes and streams in New England, Canada, and Scandinavia and to a decline of fish populations elsewhere.

The material effects of atmospheric acid are evident in damage to buildings and monuments. The acid etches and corrodes many building materials, including marble, limestone, steel, and bronze. Worldwide, tens of thousands of sculptures, buildings, and other structures are slowly being dissolved by acid rain.

Ironically, the dramatic increase in acid precipitation in recent decades is partly the result of an effort to curb air pollution. The U.S. Clean Air Act of 1970 restricted the deposit of specific pollutants over the surrounding countryside and set standards only for ground-level air quality. In order to keep air in local communities clean enough to meet the air-quality standards, industries and power plants built ever-taller smokestacks that discharge sulfur dioxide and other pollutants into the upper atmosphere. Stacks 1,000 feet (305 m) high are now a common sight; previously, stacks 200 to 300 feet (60 to 90 m) high were the norm. The situation is not unlike disposing of garbage by throwing it over your backyard fence. It still comes down, but not in your yard. Unfortunately, the farther and higher the noxious emissions go, the more time they have in which to form acids by combining with other atmospheric components and moisture; thus, the taller stacks have directly aggravated the acid rain problem.

Much of the responsibility for solving the acid rain problem lies with industry and government, but individuals can help by reducing their demand for energy. Various ways of conserving energy are discussed in the next section, which deals with the greenhouse effect, but they also apply here. The solution to both the greenhouse effect and acid rain is to reduce the emission of pollutants into the atmosphere. We can do that by using less energy and by using it as efficiently as possible.

Response: Stretch Your Gas Mileage

Motor vehicles produce from one-third to one-half of nitrogen oxide emissions in North America. That's why individuals can help stop the acid rain problem. The less gas we use, and the cleaner the gas, the fewer nitrogen oxide particles will be spewed from tailpipes. Of course, we would all benefit by driving less, but if you must use a car, here are some ways to make sure its impact on the environment is as benign as possible.

1. If you're considering buying a new car, get one that is fuel-efficient— that gets at least the average number of miles per gallon, or mpg (about 28), but you can buy cars that get 40 or 50 mpg. The more miles you get per gallon, the fewer emissions your car will produce.

2. Consider a manual as opposed to an automatic transmission. You'll use less gas.

3. Make your car more fuel efficient by having it tuned up every 5,000 to 10,000 miles (8,000 to 16,000 km). Make sure the fuel filters and spark plugs are clean. Change the oil regularly. You'll increase the fuel economy of your car by as much as 15%.

4. Check once a month to make sure your tires are inflated to the maximum pressure intended by the manufacturer. It is estimated that about half of American cars have underinflated tires. Because underinflation increases resistance, it wastes gas.

5. If you need new tires, check out steel-belted radials, which can increase fuel efficiency by 5% to 10%.

6. Do you really need that air conditioner? It uses gas even when it's not running, because it adds to the weight of the car. When it is running, it will use as much as a gallon of gas per tankful, increasing your fuel consumption anywhere from 10 to 20%. At high speeds, however, roll up your windows and use the air conditioning. You'll use less gas than driving with the windows open, which creates drag and wastes gas.

7. Try to reduce the number of car trips you make. Combine several errands into a multipurpose trip. Avoid making many short, single-purpose trips.

8. Idling wastes gas. If your car is properly tuned, you shouldn't have to warm it up for several minutes after starting the car. You should be able to wait just a few seconds for the oil to circulate, then drive away. Don't leave the car idling while you're stuck at a train crossing or waiting for a passenger. It takes less gas to restart the car than it does to let it run for more than 60 seconds.

9. Avoid jerky starts and stops, which waste fuel. Accelerate and decelerate smoothly and gradually. Minimize braking by slowing down as you

approach a red light. It may turn green before you have to come to a complete stop. Remember, it takes more gas to move the car from a full stop than from even a very slow speed.

10. On the highway, try to maintain a steady speed. Fifty-five miles per hour is best. You'll use about 20% more gas at 70 mph than at 55 mph.

Response: Fill Up Wisely

The way you fill up your car affects the amount of nitrogen oxides the car produces. The fewer emissions, the less acid rain.

If you have an old (pre-1975) "gas guzzler" that uses leaded fuel, consider getting rid of it. Leaded gas is laden with nitrogen oxides. Even if you can't afford to trade it in, you may be able to use unleaded gas or a mixture of leaded and unleaded high octane. Check with your mechanic to be sure.

If you have a car that uses unleaded fuel, be sure to fill it with unleaded. Leaded gas can ruin a newer car's catalytic converter, which is designed to reduce nitrogen oxide emissions.

Vapor controls (the plastic hoods on gas pump nozzles that are now required in many states) are there to prevent vapors from escaping into the atmosphere. Don't pull them back; the vapors help create smog.

Don't "top off" the gas tank. Stop when the nozzle clicks off. Topping off will force gas fumes into the atmosphere.

Cars don't have to run on a petroleum-based fuel. If you live in an area where alternative "clean" fuels are available, try one of them. Fuels such as ethanol, methanol, and natural gas burn more cleanly that conventional gasoline and produce fewer nitrogen oxide emissions. Although methanol and natural gas come from nonrenewable resources, ethanol can be made from a variety of renewable ones, such as corn, sugarcane, or grains. Gasohol, a blend of ethanol and gasoline, can be used in any vehicle, without engine modifications.

Finally, don't buy a car that uses diesel fuel. Even though diesel engines conserve fuel, they produce far more pollutants than do other cars and won't meet emission standards in states that have them.

CHALLENGE: THE GREENHOUSE EFFECT

Unless you're a virtual hermit, you have undoubtedly heard about the greenhouse effect. But unless you're a science major or follow the news closely, you probably are uncertain what all the fuss is about. So here, to resolve that uncertainty, we present "Everything you ever wanted to know about the greenhouse effect but were afraid to ask."

Q: What is the greenhouse effect?

A: The theory of the greenhouse effect is that certain gases concentrate in the atmosphere, where they function as an insulating barrier, absorbing infrared radiation that would otherwise be reflected back into the upper atmosphere. In other words, like glass in a greenhouse, the gases let in incoming solar radiation but retard its reradiation back into space. The greenhouse effect could cause a gradual warming of both the earth's surface and the lower atmosphere. Higher temperatures would have significant impacts on the earth's ecosystems.

Q: What gases are we talking about?

A: Carbon dioxide (CO_2), created mostly by burning fossil fuels, is the most plentiful of the gases and is thought to be responsible for about half of the warming. Each year, we send billions of tons of CO_2 into the atmosphere, only half of which is absorbed by the oceans and forests. Burning of the tropical rain forests not only adds to the emissions but also means the loss of trees that naturally absorb CO_2.

Equally important, taken together, is the accumulation of three other types of gases: (1) **methane,** from natural gas and coal mining, agriculture and livestock, swamps and landfills; (2) **nitrous oxides,** from motor vehicles, industry, and chemical fertilizers; and (3) **chlorofluorocarbons** (CFCs) and **halons,** widely used industrial chemicals. Although these gases may be present in small amounts, some of them trap heat thousands of times more effectively than does CO_2. Fluorocarbon 12, for example, has 20,000 times the capacity of CO_2 to trap heat, and fluorocarbon 11 has 17,500 times the capacity of CO_2. Even methane is thirty times more potent than CO_2 in trapping heat close to the earth.

Q: What evidence is there that these gases are accumulating in the atmosphere?

A: We know that during the last 250 years the concentration of CO_2 in the atmosphere has risen from about 274 parts per million (ppm) to over 360 ppm. (Just since 1958, concentrations of CO_2 have increased from 315 ppm.) This increase is one of the effects of the Industrial Revolution. The methane concentration in the lower atmosphere has *already* more than doubled from its preindustrial level (from 650 parts per billion to 1,700) and is currently increasing by just over 1% per year. The carbon monoxide concentration also seems to be increasing at a rate of slightly over 1% per year.

Q: Is the world really warming up? If not, why do they say it will?

A: It's too early to tell if the earth is warming up. When CO_2 concentrations reach about 550 ppm (double pre-Industrial Revolution levels), average annual global temperatures are expected to rise by 2° to 7°F (.4° to 4°C). Predictions of when this doubling will occur vary. The year 2050 is commonly cited, but because gases other than CO_2 are contributing to the greenhouse effect, the warming may occur as early as 2030.

Proponents of the greenhouse-effect theory believe that human activity has already put enough of the various gases into the atmosphere to cause a significant rise in temperature in the next century. They contend that some warming is inevitable even if all emissions were to stop today, because the greenhouse gases are already in the atmosphere.

Q: What difference does it make if temperatures rise by a few degrees?

A: Researchers have developed various mathematical models to simulate the effect of greenhouses gases on the earth. The warming will not be uniform. It will be greater at higher latitudes than in equatorial regions, and it will produce significant changes in sea level, precipitation, and vegetation. The sea level is expected to rise 1 to 4 feet (0.3 to 1.2 m) as a result of ice-cap and glacial melting and from **thermal expansion** of the water (water expands as its temperature increases). Most coastal marshes and swamps would be inundated by salt water; coastal erosion would increase. Water quality would decline as aquifers became polluted by salt. Such low-lying regions as the North American Gulf Coast, the Netherlands, the Nile Delta, Bangladesh, and much of Southeast Asia could lose substantial amounts of land. Many major ports might be flooded.

Warming of lakes and oceans would speed evaporation, causing more active convection currents in the atmosphere and thus fiercer storms. Important regional changes in precipitation would occur, with some areas receiving more precipitation, others less. Polar and equatorial regions might get heavier rainfall, and the mid-latitudes might become drier.

Changes in temperature and precipitation would affect soils and vegetation. The composition of forests would change as some areas became less favorable for certain species of plants and more hospitable to others. Hotter, drier weather would reduce crop yields in some areas, such as the corn and wheat belts of the Midwest. Conversely, more northerly agricultural regions, such as parts of Canada and Russia, might become more productive.

Q: Don't some scientists dispute the greenhouse effect?

A: Yes. Some argue that global temperatures might stabilize or even decrease as the concentration of greenhouse gases increases. A hotter atmosphere, they say, would increase evaporation, sending up more water vapor that could condense into clouds. The increased cloud cover might reflect so much sunlight that it would slow the rate at which the earth would be heated. Others contend that the increased evaporation would produce more rainfall. As it fell, the rain would cool the land and subsequently cool the air over the land.

Finally, some researchers believe that the geological record shows that large fluctuations in global temperature have always occurred independently of human activity, never as a result of it. These fluctuations are caused by such unpredictable events as variations in solar radiation, shifts in the earth's orbit and in ocean currents, meteoric activity, and volcanic eruptions.

Q: Why don't we just wait twenty or thirty years until we see what's going to happen?

A: If those who believe in global warming and its consequences prove to be correct, the longer we wait, the worse the situation will become. If we continue to spew billions of pounds of CO_2 and other greenhouse gases into the atmosphere each year, we help guarantee that the theory will become reality.

Furthermore, reducing energy consumption makes sense *regardless* of concern about the greenhouse effect. Decreasing the amount of coal, oil, and natural gas we burn saves money, saves resources, and will extend the life of precious fossil fuel reserves. It will reduce smog and acid rain. Investing in energy efficiency will give us cleaner air and water, save money on utility costs, and reduce our dependence on foreign oil.

Q: What can I do?

A: Read pages 61–75.

Response: Use Energy Efficiently

The fastest and cheapest way to combat the greenhouse effect is to conserve energy—and the fastest and cheapest way to conserve energy is to use it efficiently. Obviously, government and industry can do much to reduce our dependence on fossil fuels, but the individual consumer can do a lot to reduce demand. Our dormitories, apartments, and houses are filled with energy-using appliances. The pages that follow suggest ways to make them as energy efficient as possible. As is the case with recycling, no one expects you to utilize all of these suggestions, but if you adopt even a few of them, you'll be doing your bit to stave off global warming.

Before we begin, take a look around your room or apartment. How many appliances use electricity? Are all of them necessary? Of course you want hot water and electric lights, but do you really need an electric can opener or pencil sharpener? An electric knife or toothbrush? None of them uses much energy, but the proliferation of electric gadgets is a symbol of our energy profligacy. We are used to wasting energy because it has been relatively cheap. Perhaps the ultimate symbol of that frame of mind is the television set with standby power for remote control. According to the Rocky Mountain Institute, most television sets now use from 1.5 to 1.8 watts of power even when they're turned off. It takes the equivalent of one power plant to keep all those TV sets powered when they're not even on!

TV sets aside, it's the major appliances in your living quarters that use the most energy. By major appliances we mean things like the hot water heater and refrigerator. There are two important facts to keep in mind about all of them. First, they vary greatly in the amount of energy they use depending on how they are built. An inefficient water heater or fridge will use twice as much electricity as an energy-efficient one. Second, you can reduce the amount of energy your appliances use, even if you are stuck with old, inefficient ones, by following some of the suggestions given in this section.

When you find yourself in the market for a new appliance, check the EnergyGuide label on it. By law, these stickers now appear on all new clothes washers, dishwashers, freezers, refrigerators, room air conditioners, and water heaters sold in the United States. The label indicates how much it will cost you to run the appliance for one year. A line scale indicates how the model you're looking at compares with others of the same size as far as energy use.

Some appliances also feature an Energy Star emblem, which indicates they are energy efficient. The Energy Star label program was devised, in 1993, by the U.S. Department of Energy and Environmental Protection Agency for computers and other home-office equipment. Recently, it has been expanded to include four types of appliances—washing machines, refrigerators, dishwashers, and room air conditioners. Any appliance displaying the Energy Star label must exceed federal energy-efficiency standards by at least 13%. Many refrigerator models, for example, exceed government energy standards by 20% or more. Over the lifetime of the refrigerator, you would save hundreds of dollars in operating costs by purchasing one of these models.

New appliances sold in Canada have an EnerGuide label that indicates the amount of energy the appliance uses per month, expressed in kilowatt hours (kWh). The lower the number, the more efficient the appliance. A freezer that uses 45 kWh per month is twice as efficient as one that uses 90 kWh.

Response: Make Your Water Heater More Energy Efficient

In most homes, the water heater is second only to the furnace in the amount of energy it uses. There are three ways you can make it more energy efficient:

1. Set the water heater to 120°F (about 50°C). This will give you adequately hot water. If the heater doesn't have temperature markings, experiment with a setting somewhere between "warm" and "hot."

2. Insulate it with a water heater blanket. These insulated paddings, made to fit different sizes of heaters, are readily available. You'll probably recoup the cost of the blanket in the first year of operation. You can also insulate the water pipes that lead away from the heater.

3. If you're going to be gone for three or more days, turn the heater to the "vacation" or "low" setting. This will keep just the pilot light lit. Otherwise, the heater will keep heating the water to a constant temperature even if no one is using it, which is not only an expensive proposition but a needless waste of energy.

When you're in the market for a new heater, keep the following points in mind:

- In the long run, it will pay you to buy the most energy-efficient model you can afford.

- Gas and oil-fired heaters use less energy than do electric heaters.

- Consider purchasing either a demand-type water heater or a solar heater. **Demand-type water heaters** do not have a water storage tank. They heat water only when you need it. When you turn on a hot water tap, the water will pass through a heating element and exit through the tap. **Solar water heaters** use energy from the sun to heat the water. They will work in all but the cloudiest of climates, and millions are now in use around the world.

Response: Buy Compact Fluorescent Bulbs

"Pay $15 for a lightbulb? You've got to be kidding!" That might be your first reaction to the suggestion that you buy a compact fluorescent lightbulb. In the long run, however, you'd be saving money—and protecting the environment at the same time.

Lighting accounts for about 25% of the electricity used in America. Generating that electricity at a coal-burning power plant results in massive amounts of carbon dioxide (and smaller amounts of sulfur dioxide) being released into the atmosphere. CO_2, as you know, is one of the chief gases responsible for the greenhouse effect; SO_2 contributes to acid rain. Using nuclear energy to generate the electricity results in the production of radioactive wastes such as strontium-90 and cesium-137.

Many power plants could be retired if people switched from standard incandescent lightbulbs to compact fluorescents. These new bulbs come in a variety of shapes and sizes. They screw into regular lightbulb sockets and are "color-enhanced" so that the light they emit is just like that from incandescent bulbs. They don't give off the eerie white light you may associate with fluorescent fixtures.

Most important, the bulbs are extremely energy-efficient, using only one-fourth to one-fifth of the energy of ordinary bulbs. Thus a 15-watt compact fluorescent gives the same amount of light as a 60-watt bulb, and an 18-watt compact fluorescent can be substituted for a 75-watt bulb. Although the initial cost of the bulbs is high, they last ten to twenty times longer than incandescents. That's why, in the long run, you'll save money two ways—in lowered electricity charges and in less cost for replacing bulbs. It is also why many utility companies are selling compact fluorescents at a substantial discount, to encourage their use.

In addition to reducing the demand for electricity, compact fluorescents will reduce the output of pollutants. If every household in America installed a single compact fluorescent, we would save the amount of energy a nuclear power plant generates in a year. If households used only fluorescents, approximately forty large coal-burning power plants could be shut down, preventing the emission of millions of tons of carbon dioxide.

Because the compact fluorescents are expensive, you probably won't want to buy many of them at once. But you might want to use them in the light fixtures you use the most, or adopt the strategy of replacing an incandescent with a compact fluorescent every time one burns out. If you live in a dormitory or apartment house and don't have to pay for your own electricity, suggest to the managers that they consider installing the fluorescents in public areas or subsidizing your purchase of them.

If even the cost of a single compact fluorescent is too high for you at this time, here are other ways to reduce your lighting bills:

- If you're going to be gone for more than a couple of minutes, turn off the lights when you leave a room. Contrary to many people's belief, it does *not* take more energy to turn off and then turn on lights than it does to leave them lit.

- Dust light bulbs occasionally and remove dead insects from light fixtures. Dust and other debris reduce the light a bulb gives off and waste electricity.

- Install dimmer switches for incandescent bulbs. They save energy and extend the life of the bulb. Note: dimmer switches should *not* be used with compact fluorescents.

- Don't use higher wattage bulbs than necessary. Hallways and closets, for example, may not require strong lighting.

- On the other hand, in areas where you need strong light, replace multiple-bulb fixtures with those that use a single bulb. One strong bulb is more efficient than several weaker ones. For example, a single 100-watt incandescent bulb gives off as much light as two 60-watt bulbs or seven 25-watt bulbs but uses less energy.

- Use solar-powered lights where possible. Although they are expensive, they cost nothing to operate and are ideal for lighting up gardens and walkways.

Response: Reduce the Cost of Running Your Refrigerator

You might be surprised to learn that refrigerating food is a relatively expensive proposition. It accounts for about 7% of the total electricity used in the United States. Approximately one-fifth of the electric bill of an average household is due to refrigerators and freezers.

There are a number of steps you can take to reduce the cost of running refrigerators and freezers:

1. As a parent undoubtedly pleaded with you when you were a child, decide what you want before you open the refrigerator or freezer door. The cold air escapes when the door is opened, and the appliance has to work overtime to cool the air again. (Of course, if you're going to keep it open for one minute, you might as well leave it open for five; the cold air has already left.)

2. Let hot foods cool down to room temperature before putting them in the refrigerator or freezer.

3. Instead of defrosting frozen foods on the counter or in the microwave, try to allow enough time for them to thaw in the refrigerator. The frozen food will help keep the air in the refrigerator cold, thus cutting down on the amount of electricity needed to operate it.

4. If you have a manual defrost model refrigerator or freezer, don't let the ice build up. Defrost the appliance every time the ice is more than 1/4 inch (0.5 cm) thick. Otherwise, the motor has to run longer to maintain a cool temperature.

5. If you find that you have to defrost often, the door seals on your refrigerator or freezer may need to be replaced. One way to check this is to place a piece of paper halfway inside the door. Hold the other half while you shut the door. If you can pull the paper out easily, either the door is warped or you need new rubber seals.

6. Clean the dust off the condenser coils on the back of the refrigerator or freezer at least once a year. A buildup of dust reduces the operating efficiency of the appliance. Pull the machine away from the wall, unplug it, and then clean the coils with a brush or vacuum cleaner.

When it's time to buy a new refrigerator or freezer, keep these points in mind:

- Check the EnergyGuide labels (see pages 61–62) on models in your price range, then buy the most energy-efficient model you can afford. The most efficient ones use about half the amount of electricity than the least efficient ones do.

- Side-by-side refrigerator-freezer combinations use about 15% to 25% more energy than do models that have the freezer on top.

- Chest-type (top-loading) freezers tend to be more energy efficient than upright models.

Response: Use Clothes Washers and Dryers Efficiently

Whether you own your own washer and dryer or use the ones provided in an apartment house or dormitory, there are a number of ways you can cut down on the amount of energy the appliances use.

Clothes washers use energy both to heat the water and to run. Here are some energy-saving suggestions:

- In general, it makes sense to use the washer only when you have a full load. If you have less than a full load, and if the machine has a water level control, adjust the setting to "low" or "medium."

- Many washers have a "light" or "gentle" cycle that takes less time and doesn't use as much water as the "normal" or "regular" cycle. The light cycle is sufficient to clean most clothes.

- Since much of the energy goes for heating the water, try washing with cold or warm water instead of hot. You'll probably find the results perfectly satisfactory, especially for dark clothes. And you can use a cold water rinse for all clothes, white or dark.

The best way to save energy when drying your clothes is to let them air dry. But not everyone lives in the kind of climate (or housing) where that's possible, so when you use a clothes dryer, remember these three tips:

- Make sure the exhaust hose is as short and straight as possible, so that the air can circulate unobstructed.

- Clean the lint filter every time you use the dryer. A clean filter helps the air circulate effectively.

- Don't run the dryer an extra thirty minutes just to get a few heavy things such as towels or sweatshirts dry. Remove them from the machine while they're still damp and hang them over a towel bar or clothesline to finish drying.

Response: Save Energy When You Use a Dishwasher

Like clothes washers, automatic dishwashers use energy both to heat the water and to operate. And like clothes dryers, they also use energy to heat the air to dry what's inside. As a result, you can use the same types of techniques to save energy when using a dishwasher as you can with clothes washers and dryers.

■ Run the dishwasher only when it's full. If you rinse the dishes, silverware, pots, and pans before loading them into the machine, they can sit there for several days without harm.

■ Most machines have a "light" wash cycle that takes less time to complete than does the normal cycle. It uses less hot water and less energy. You should be able to use it for nearly all of your dishwashing needs.

■ If your water heater is set at 120°F (about 50°C), you should use a dishwasher that has a booster heater. The heater raises the temperature of the water entering the machine to 140°F (60°C), the temperature recommended to thoroughly clean greasy dishes. If you're not sure whether the machine you use has a booster heater, check the manual that comes with the washer or call an appliance dealer to find out.

■ Be sure to let the dishes air dry. This will save 20% to 30% of the electricity used for a complete cycle. If your machine doesn't have an "air dry" option, simply open the machine and pull out the racks. Opening the washer will stop the action, and the dishes will dry in short order.

In many kitchens the dishwasher is next to the refrigerator. If this is the case in your kitchen, slide a board or a piece of fiberglass or other insulating material between the appliances. This will help keep the heat generated by the dishwasher from warming the air next to the refrigerator.

Response: Cook with Less Energy

There are a number of ways you can cut down on the amount of energy you use when you cook:

- The first rule of thumb is to use the smallest appliance you can for cooking something. If you're broiling a single hamburger or warming up one TV dinner, use a toaster oven if you have one. You'll use less energy than if you heat an entire oven for that task. If you're cooking on top of the stove, use a pot that covers the heating element. If the bottom of the pot is smaller than the element, you're wasting energy by heating the air.

- Microwave ovens and pressure cookers are energy efficient. A microwave oven uses only half the energy that a conventional oven does to cook something. Not only is it smaller than the typical oven, but it heats just the food, not the container. And because the microwave makes it easy to heat leftovers, you're less likely to throw them out. A pressure cooker can cook a roast or chicken in about one-third the time it takes a conventional oven to do so.

- For stove-top cooking, use lids for your pots. They help keep the heat in the pot, which reduces cooking time.

- Use glass rather than metal baking dishes. They transfer heat more efficiently than do metal pans, which is why you can reduce the oven temperature 25°F (14°C) when baking a cake in a glass pan. Try to resist the temptation to open the oven door while something is cooking. You'll be letting in cooler air.

When you or someone close to you is in the market for a new stove/oven, remember these facts:

- Gas ranges, though initially more expensive than electric ranges, require much less energy to operate and will save you money in the long run.

- Gas ranges are now available with electronic ignition systems instead of pilot lights. A pilot light burns continuously, wasting gas needlessly. An electronic ignition system can cut your gas bills for cooking by as much as half.

- Convection ovens have a fan that circulates the air throughout the oven. Because the fan eliminates hot and cold spots, cooking times are shortened.

Response: Keep Your Cool

Air conditioners are marvelous conveniences, as you know if you've ever sweltered through a hot, humid summer without one. Unfortunately, manufacturers use chlorofluorocarbons (CFCs) as the cooling fluid in air conditioners, and the electricity to run them releases CO_2 into the air. CFCs and CO_2 are greenhouse gases. While we don't ask you to do without air conditioning if you live in a climate that demands it, we do request that you use it both wisely and efficiently. You'll be rewarded with lower energy consumption and lower electricity bills.

Getting Along without Air Conditioning   The less you use the air conditioner, the better for the environment. Take precautionary steps to keep summer heat from building up in your room, apartment, or house.

- Shut your windows and outside doors during the hottest part of the day. Draw the curtains or blinds, pull down the shades. Consider mounting awnings or shutters on the windows. During cooler hours, or at night, open the windows and let in fresh air.

- Use as few heat-generating appliances during the day as possible. Do your baking and clothes washing and drying in the early morning or during the evening.

- Consider alternatives to the air conditioner. Ceiling fans use only a fraction of the electricity that air conditioners do, so they cost considerably less to run. Ventilating or exhaust fans, particularly large attic fans, can do an excellent job of cooling. They pump heat from the house to the outside. Even the exhaust fan over your stove will help vent a hot kitchen.

Wise Use of an Air Conditioner   For those times when you feel you must use an air conditioner, we offer the following suggestions for doing so as efficiently as possible:

- Keep the air-conditioning unit as cool as possible. If it is in the direct sunlight for much of the day, shade it with an awning, or plant a tree or shrub nearby.

- Set the air-conditioning thermostat as high as possible. 78°F (25°C) should be a reasonably comfortable indoor temperature.

- Keep lamps or other heat-generating appliances away from the thermostat. Otherwise, it will sense the heat and make the air conditioner run longer than necessary.

- Don't assume that the best way to cool a room or house rapidly is to turn the thermostat to a cooler than normal setting, then raise the temperature later on. All you'll do is cool the room to a lower than necessary temperature, wasting energy in the process.

- Turn off the air conditioner if you'll be gone for more than a couple of hours. If you often forget to do so, use a timer to control the air conditioner. Programmable models let you select several different temperatures for different times during the day and night. You can set the air conditioner for one temperature while you're at home, a higher temperature while you're normally away, and then have it cool the house again just before you get home.

- Clean or replace the air conditioner filter every month or two during the summer. A dirty filter means the fan has to run longer to move the air, so you're using more electricity than necessary.

- Try running the air conditioner in conjunction with one or more fans placed so as to blow the cool air into adjacent rooms.

Response: Conserve Energy While Keeping Warm

For most of us, keeping warm in the winter is an expensive proposition. Home heating typically uses more energy than any other single function, including heating water or providing light. Heating systems also contribute carbon dioxide and sulfur and nitrogen oxides to the atmosphere, gases that are responsible for global warming and acid rain.

Unfortunately, much of the energy used to heat homes is wasted. Some of it leaks out of the house, and some of it need never have been used to heat the air in the first place. In this section we suggest some simple ways to conserve energy and to use it as efficiently as possible.

Insulate Your Home   Whether you live in an older building or a relatively new one, it may be poorly insulated. That means you'll be losing large amounts of heat through the walls and attic. If you live in a cold climate, one way to check to see if heat is lost through the roof is to look at it a day or two after a heavy snowfall. If large patches of snow have already melted off, heat is probably escaping.

The standard procedure for insulating walls is to blow in insulation, not something you're likely to do unless you're a homeowner. But insulating an attic or crawl space is relatively easy. Look to see if there is any insulation at all. If there is, it should be at least 3 inches (8 cm) thick and be evenly distributed, with no spaces between the batts of insulation. Should you need to add more, follow the instructions that come with the insulation. To prevent fires, insulation should be kept at least 3 inches (8 cm) from chimneys, lights, and other heat-producing fixtures.

Weatherize Your Home   A surprising amount of heat escapes through doors and windows, cracks in the foundation, and even electrical outlets. You can see where heat is being lost by slowly moving a lighted candle around door and window frames on a cool, windy day. If the flame flickers, cold air is coming into the house (and warm air is leaving it).

Windows

- The most energy-efficient windows have double or even triple panes, with a thin layer of air between the panes. If you don't have these, make sure your windows have storm windows. If you don't want to invest in storm windows, you can fasten heavy-duty, clear plastic sheets over the windows from the inside. An alternative, though less satisfactory, is a thin plastic film that is applied to the window frame with double-stick tape.

- Remove window screens during the winter; they block some of the sun's rays.

- Ideally, a window air conditioner should also be removed before winter sets in. If that isn't practical, insulate it from the outside.

- Caulk and weather-strip all windows. Caulk is a semi-liquid material that is placed around permanent joints, such as where the window frame meets the wall. It's often applied with a caulking gun. Weather stripping, which comes in several different forms, is put around movable joints—for example, where the window meets the sill.

- Finally, insulated or "thermal" shades and drapes can help save energy. Open them during the day, especially on windows that receive sunlight, and close them at night.

Front and Back Doors

- Install a storm door if the house doesn't have one.

- Caulk and weather-strip.

- If a lot of cold air comes in under the door, a "snake" is useful. This is a fabric tube stuffed with sand, beans, or small pellets. For a more permanent solution, install a door threshold seal. There are many different types. Some mount to the door itself, others to the floor beneath the door.

Electrical Outlets on Exterior Walls

- Outlets on exterior walls can be insulated with special foam pads that fit between the cover plate and the electrical box. Turn off the power before unscrewing the cover plate.

- If certain sockets are never used, block the holes with plastic inserts (the kind intended to make the socket childproof).

Outside the House

- Caulk any cracks.

- Caulk around any openings, such as where plumbing or telephone wires enter the building.

The Fireplace   Although fires are lovely, most are not efficient at space heating. Furthermore, burning wood releases carbon dioxide, which contributes to the greenhouse effect. Here are some suggestions for making fireplaces more energy efficient.

- Install glass doors, which are not only decorative but help cut down on the loss of warm air through the chimney.

- Be sure the damper is closed except when you're using the fireplace.

- If you never use the fireplace, close it off with a wood, glass, or metal screen. Even when the damper is closed, some air will escape up the chimney.

Radiators   If you have radiators, there are several steps you can take to make sure they function as well as possible.

- Make sure radiators are clean and working properly. They need to be "bled" occasionally and may need new valves. If a *hot-water* radiator doesn't heat up along its entire length, air is probably trapped inside and the radiator needs bleeding. Unscrew the bleeder valve at the top of the radiator and hold a pan under the valve. Air should escape, then hot water. When the water is flowing freely, close the valve. A *steam* radiator does not require bleeding; if it doesn't heat fully, the vent may be clogged. Shut off the radiator and when it is cool, remove the vent. You can try shaking the vent to dislodge rust particles or boiling the vent in vinegar for 20 minutes to dissolve mineral deposits. If these don't work, buy a new valve.
- An aluminum reflector placed behind a radiator will reflect heat back into the room.
- A tray of water placed on top of the radiator will help humidify the room.

The Thermostat

- If there is a window near the thermostat, make sure it is closed tightly, caulked, and weather-stripped. Otherwise, the thermostat will sense the cold air and keep the furnace working.
- Turn down the thermostat if you're going to be gone for several hours and also when you go to bed at night.
- A programmable thermostat can be set so as to turn your furnace down at night, up in the morning, down when you leave for the day, and up again before you return in the afternoon. Although such thermostats are costly ($40-60), they could save as much as 20% of your annual heating bill.

Response: Drive a Fuel-Efficient Car

As you know, burning fossil fuels, including gasoline, produces carbon dioxide, one of the principal greenhouse gases. Motor vehicles are a major contributor to the greenhouse effect, accounting for about one-third of the U.S. total carbon dioxide output. The problem is getting worse, not better, as the number of cars on the road keeps increasing. Annual vehicle miles traveled in the U.S. have more than doubled since 1970. The widely popular sport utility vehicles, pickup trucks, and vans are among the least fuel-efficient vehicles.

For every gallon of gas a car burns, it contributes almost 20 pounds (9 kg) of carbon dioxide to the atmosphere. That might seem to defy logic. How could a gallon of gas, which weighs about 6 pounds (2.7 kg), produce 20 pounds of CO_2? The same way that you can gain a pound after eating a few ounces of fudge. Burning the gas releases about 5 pounds (2.3 kg) of carbon. In the atmosphere, each carbon atom combines with two oxygen atoms to form CO_2—which weighs almost four times as much as carbon does by itself.

Making fuel efficiency a national priority would help slow the rate of global warming. The less gas a car burns, the less carbon it will spew into the atmosphere and the better off we will be. As a result of the OPEC oil embargo of 1973-74, the U.S. Congress required automobile manufacturers to increase gas mileage. That legislation had a significant impact. The average fuel efficiency of cars on the road rose from 13 mpg in 1973 to about 20 mpg today. New cars average almost 28 mpg, but it's possible to do much better. Many car models get over 35 mpg even in city driving.

These are some of the benefits of increasing the number of miles per gallon that our cars get:

- Less smog.
- Less acid rain.
- Less CO_2, a greenhouse gas.
- Lower costs for gas, since we don't have to fill up as often.
- Less pressure to drill for oil in sensitive environments, such as offshore oil fields or Alaska's national wilderness areas.
- Less dependence on importing oil from abroad.

What You Can Do   If you already own a car, make it as fuel efficient as possible. See pages 55-56.

When you're in the market for a car, check the fuel efficiency ratings of the models you're interested in. The federal government publishes these each fall. Information is also available in *Consumer Reports* and automotive magazines. New cars have this information on their stickers.

CHALLENGE: DEPLETION OF THE OZONE LAYER

Ozone, which is a noxious pollutant near the ground, is an essential chemical in the stratosphere. There, some 6 to 15 miles (10-24 km) above the ground, it forms a protective blanket, the ozone layer, which shields all forms of life on earth from overexposure to lethal ultraviolet (UV) radiation from the sun.

Emissions from a variety of chemicals are destroying the ozone layer. The primary culprit is the family of synthetic chemicals developed in the 1920s and known as chlorofluorocarbons, or CFCs, combinations of carbon, fluorine, and chlorine atoms. These were initially hailed as perfect chemicals because they are nontoxic, noncorrosive, nonflammable, long-lasting, don't accumulate at ground level, and are relatively easy to manufacture.

CFCs are found in hundreds of products around the house. They are coolants for refrigerators, freezers, and air conditioners. They are aerosol spray propellants, and a component in foam packaging, home insulation, and upholstery. In liquefied form, they are used to sterilize surgical equipment and to clean computer chips and other microelectronic equipment.

Also implicated in the depletion of the ozone layer are halons, used in fire extinguishers, and carbon tetrachloride and methyl chloroform, used as solvents and cleaning agents. CFCs, however, are by far the most important. More than 2 billion pounds (900 million kg) of the chemicals are manufactured each year.

After the gases are released into the air, they rise through the lower atmosphere and, after a period of seven to fifteen years, reach the stratosphere. There, UV radiation breaks the molecules apart, producing free chlorine atoms. (Halons release bromine atoms.) Over time, a single one of these atoms can destroy tens of thousands (if not a potentially infinite number) of ozone molecules. Bromine is even more destructive than chlorine.

Every Southern Hemisphere spring, beginning in late August, the atmosphere over the Antarctic loses more and more ozone. In 1985, researchers discovered what is popularly termed a "hole" as big as the continental United States in the ozone layer over Antarctica, extending northward as far as populated areas of South America. It is not truly a hole, but depletion in the ozone layer is as much as 60%. The hole disappears in late November, when the winds change and the ozone-deficient air mixes with the surrounding atmosphere. A less dramatic but still serious depletion of the ozone shield occurs over the North Pole.

A depleted ozone layer allows more UV radiation to reach the earth's surface. Although the exact consequences of that increase won't be known for years, it is almost certain to cause a dramatic rise in the incidence of skin cancers and eye cataracts. Some fear it may also damage the immune system of humans and other animals by impairing the cells that fight viral infections and parasitic disease. Because UV radiation also causes cell and tissue damage in plants, it is likely to reduce agricultural production severely. The most serious damage may occur in oceans. Increased amounts of UV adversely affect the photosynthesis and metabolism of the microscopic plants called phytoplankton that flourish just below the surface of the Antarctic Ocean. Phytoplankton form the base of the marine food chain and play a central role in the earth's carbon dioxide cycle. A loss in plankton will affect the entire marine food web, including fish, marine mammals, and birds.

The production of CFCs is being phased out under the Montreal Protocol on the Depletion of the Ozone Layer of 1987, an international agreement endorsed by almost 150 countries. The treaty required developed countries to stop production of CFCs by 1996; developing countries have an extra ten years to cease production. Even if all countries eventually comply with the Montreal Protocol, however, past emissions will continue to cause ozone degradation for years to come. The two most widely used forms of CFCs stay in the stratosphere for up to one hundred years. That means that some of the first CFCs ever produced are still in the stratosphere, breaking down ozone molecules. So even if use of the chemicals were to stop tomorrow, it would take the planet a century to replenish the ozone already lost.

Response: Keep CFCs Where They Belong

Although the United States and Canada halted production of CFCs in 1996, appliance manufacturers, repair firms, and some other companies are allowed to recycle and reuse existing stocks of CFCs. Furthermore, businesses can still sell older CFC-using models of refrigerators, air conditioners, and other appliances. Until all such products are free of CFCs, do what you can: Don't let the CFCs be released into the air, and stop buying products that contain them.

Car Air Conditioners   Automobile manufacturers are among the largest users of CFCs in the United States and Canada. Air conditioners have now become virtually standard equipment in cars. Unfortunately, these air conditioners tend to leak, and hence to require several refills of coolant over their lifetime. The standard practice in servicing car air conditioners has been to vent the old refrigerant, releasing the CFCs into the air, and then replace it with new coolant.

An increasing number of repair shops now have equipment that enables operators to capture and recycle the refrigerant. Proper recycling doesn't put any CFCs into the atmosphere. Be sure to patronize one of these repair shops when your air conditioning unit needs servicing.

You may do your own car maintenance. If so, don't buy a can of Freon (the Du Pont brand name for CFCs) refrigerant to top up the cooling system. It's better to have the leak properly repaired.

Refrigerators and Freezers   Manufacturers are now producing CFC-free refrigerators and freezers, but you probably have an older model that contains CFCs. Make sure that when one of these appliances is serviced, the technician uses special equipment to capture and recycle the coolant.

Also, when you're defrosting a freezer unit, be sure not to use a knife or other sharp object to dislodge the ice. If you puncture a coil, CFCs will escape into the air.

Insulation   Don't buy building insulation that is made with CFCs. Good alternatives include cellulose, fiberglass, and gypsum. If you want foam insulation, check the label; non-CFC foam is manufactured.

Furniture   Don't purchase furniture that is made with CFCs. This includes foam mattresses, chairs and couches, even carpet padding. If you're not sure whether the foam was made with CFCs, read the label or ask the retailer to check with the manufacturer.

Aerosol Sprays   Although the United States and Canada banned most use of CFCs as propellants in aerosol spray cans in 1978, loopholes in the law mean that roughly 10% of aerosols still use CFCs. These aerosols include cleaning sprays for film negatives, sewing machines, VCRs and other electronic equipment, and cans that spray plastic confetti. Each time you use such a spray, you release ozone-destroying CFCs into the atmosphere. Check the label before buying any of these products. If it contains CFCs, put it back on the shelf.

Fire Extinguishers   Similarly, don't buy a home fire extinguisher that uses halons as propellants. Halons are even more destructive to the ozone layer than are CFCs. Even if you never use the fire extinguisher, the halons will eventually leak out of it into the atmosphere. Home fire extinguishers that use dry chemicals are a satisfactory alternative.

EPILOGUE

According to recent polls, three-quarters of all Americans consider themselves "environmentalists," but there is no agreement on what that term means. At one end of the spectrum are those who think being environmentally responsible means consuming as little as possible; they embrace a back-to-the-earth philosophy and a simple lifestyle that has the least possible impact on the earth. Some grow their own food, eat no meat, give up cars, even forswear the use of perfume or cologne because of their airborne emissions. At the other end of the environmental spectrum are people who think they have done their bit to clean up the environment by recycling aluminum cans.

Certainly, most people don't want to go back to a less comfortable, less convenient way of life. We want the goods and services that our technology and affluence afford us. You have to decide for yourself what being environmentally responsible means. You must set your own priorities. It might help to consider the points that follow:

- Many people believe that living in the United States or Canada imposes special obligations on us. With less than 5% of the world's population, those countries consume about 25% of the world's energy—about twice as much as Western Europe and Japan per unit of economic output. Among other things, this energy use contributes to global warming, acid rain, oil spills, and nuclear waste.

  The population biologist and ecologist Paul Ehrlich devised a simple equation to measure the environmental impact of a society:

  $$\text{Impact} = \text{Population} \times \text{Affluence} \times \text{Technology}$$

  That is, the impact of a population on the earth's ecosystems and resources is a result of the size of the population, its affluence (measured as the amount of natural resources the average person consumes), and the technology used in producing and disposing of the goods consumed. Considered this way, an American has twice the environmental impact of a Swede, 13 times that of a Brazilian, and 140 times that of a Kenyan. Ehrlich argues that instead of operating a wasteful economy, North Americans should take the lead in showing how environmental protection, economic efficiency, and human health can go together.

- As Americans, most of us have been conditioned by the dominant historical view of the relationship of human beings to the natural world. A product of religious beliefs, democratic ideals, and social and economic concepts, the Western worldview is anthropocentric (human-centered) rather than biocentric. These are some of its controlling assumptions:

· · ·humans can achieve dominion over nature;

· · ·individuals have inherent rights to own property and to develop resources;

· · ·the accumulation of wealth is a virtue;

· · ·resources have no inherent value beyond their usefulness to human beings;

· · ·science and technology can solve most serious problems.

Against these, environmentalists set competing paradigms:

· · ·humans should revere nature and be stewards of the earth;

· · ·the earth does not belong to us; we belong to the earth;

· · ·economic growth must be limited;

· · ·resources are finite and exist in a delicate balance that humans must observe;

· · ·science and technology will never enable us to circumvent natural laws or control natural processes.

The ways you choose to be environmentally responsible will be determined in part by your own view of the relationship between people and nature.

- The purpose of this book has been to provide you with easy, practical ways to mitigate your impact on the earth and to help you understand why the task is so important. But the book gains significance only when you choose to act—when you choose ways of living that protect the air, water, and land of Spaceship Earth.

- Recognize that we live in a consumer society, bombarded by advertisements urging us to buy newer, better goods and services. Our purchasing decisions and behavior have a direct effect on resource use. One of the most important things you can do is to slow your demand for products and services that contribute to the depletion of natural resources. Reject advertising-driven overconsumption: Buy less stuff.

- Recognize, too, that some actions are more important than others. Driving a well maintained, energy-efficient car is more important than recycling. But reorganizing one's life to significantly reduce dependence on a car is better yet. The choice of where to live, and the transportation and consumption choices involved in this decision, are among the actions we take in our lives with the most overall environmental impact.

- Be aware of the multiple effects that a single action can trigger. For example, planting a garden that is adapted to the local climate can save water, fertilizers, and pesticides; sequester carbon in trees and reduce runoff; provide a habitat for birds and small animals; provide food for the family; shade one's home to reduce summer air conditioning; provide a wind break to reduce winter heating; and give one a sense of connection to the earth and living things.

- Help children gain first-hand exposure to the outdoors. Many children have little chance to experience the natural world. They are consumed by television, computer games, organized sports, or other accoutrements of a high-tech society. As we hear so often, our children are our future, but it's hard for them to care about the environment if they haven't had much direct contact with it. Take children, your own or others, to the woods or a meadow, to a pond or ocean beach, and let them explore.

- Set an example for others. We all act in multiple spheres—our families, schools, workplaces, religious and social organizations, the businesses we patronize. Countless opportunities to be environmentally responsible present themselves every day. The light you turn off when you leave a room, the can you recycle, the low-phosphate detergent you use all will help lessen your impact on the earth *and* will demonstrate to others your concern for the environment. Remembering to do these things takes an effort at first, but it soon becomes automatic. Actions that seem inconsequential by themselves become important when millions of people adopt them.

- Pick an issue that you think is important and become well informed about it. If you would like to know more about any particular subject mentioned in this book, whether it be city dumps or tropical rain forests, your librarian or instructor will be able to help you investigate it further. Try to learn about different aspects of the same subject, for instance biological, social, economic, and legal considerations. You become smarter and more effective when you understand a problem from different angles.

- Let your voice be heard. Vote for candidates who demonstrate an abiding concern for the environment. Write your local, state, or federal government representatives when important issues arise. Write businesses if you think they are harming the environment. Support environmental organizations whose goals match your own, whether it is protecting endangered species, saving tropical rain forests, supporting the use of renewable energy resources, or another cause.

- Show your commitment to the place where you live by offering your time as a volunteer. By taking an interest in your community, you help shape its future.

In sum, find out what role you can play, taking into account your own interests and abilities. Whatever you like to do, you can do it for yourself and the whole world, as soon as you realize that you can make a difference. We hope this book has helped you to know it.